Business Strategies for
Satellite Systems

For a complete listing of the *Artech House Space Applications Series*,
turn to the back of this book.

Business Strategies for Satellite Systems

D. K. Sachdev

Artech House
Boston • London
www.artechhouse.com

Library of Congress Cataloging-in-Publication Data
A catalog record of this book is available from the U.S. Library of Congress.

British Library Cataloguing in Publication Data
Sachdev, D. K.
    Business strategies for satellite systems.—(Artech House space applications series)
    1. Direct broadcast satellite television—Management   2. Strategic planning
    I. Title
    384.5'544'068

ISBN 1-58053-592-5

Cover design by Gary Ragaglia

International Standard Book Number: 1-58053-592-5

10 9 8 7 6 5 4 3 2 1

To my wife, Usha

Contents

Preface

Business Strategies for Satellite Systems presents the essentials of strategy development and project management in one integrated approach as applicable to modern satellite-based systems. Throughout the book, the emphasis is on maximizing the likelihood of business success through active interaction among all major functions. The book does not claim to compete with or try to overlap several excellent books dedicated to the underlying individual areas of satellite communications, strategy development, and project management. Instead, its main objective is to combine the essentials from all of these disciplines in one integrated process. While the text is centered on modern satellite systems, it is expected that some of the principles and methodologies presented would find applications in other fields as well.

The book is sufficiently broad and generic to attract a fairly wide audience in the professional as well as academic environments. Senior executives and professional managers will find in one place treatment for all of the stages involved, starting with the vision of an enterprise or project and continuing all the way to actual implementation, with principal focus throughout on managing rather than on details. Many of the managers will no doubt recall their own experiences, perhaps with their own unique lessons to compare with those presented here. Young professionals, aspiring to become first-time managers or independent entrepreneurs on their own, will find in this book a useful compendium of the relative roles of different activities and their mutual interactions.

A good part of the material presented in this book has been used recently for teaching graduate classes in system engineering and project management. The students find these courses complement very well their other classes on theoretical and engineering aspects. As a minimum, the students and their teachers will find it convenient to use this book as a single text instead of multiple textbooks in order to cover all aspects related to satellite-based businesses.

This book reflects to a certain extent my professional experience of over 30 years in the satellite field under a variety of environments. My first serious introduction to this fascinating technology was in 1968, as a member of a three-month system study jointly with Hughes Aircraft at El Segundo, California, for a future satellite system for India. This assignment also brought me in direct contact with the legendary Harold Rosen and enabled me to learn firsthand about the myriad technical and nontechnical aspects he and his associates had to address only a few years earlier in order to successfully launch and demonstrate the very first geostationary satellite, Early Bird.

About 10 years later, a long association started with INTELSAT, initially with the responsibility to manage its research and development program. After some years of managing medium- and long-term technology topics around the world, my responsibilities were expanded to include the development of new satellite systems for Intelsat. It was during this phase, spread over a decade, that many of the concepts presented at some length in this book began to take shape. The decision processes at INTELSAT—at that time, still an international consortium—despite all of their quirks, had the unique advantage of bringing us in regular contact with literally hundreds of professionals from all over the world. Such interactions often provided excellent opportunities to watch fairly closely what works and what does not in the technology and business worlds under different external and internal environments. A concrete opportunity to test such ideas came with the responsibility to plan and develop the INTELSAT VII program in the late 1980s. With the active support of a progressive internal management, it became possible to try different organizational paradigms. As a result, a highly interactive team of several previously somewhat-isolated functions developed fairly quickly a highly flexible (and fungible) spacecraft series, as captured briefly in Chapter 6.

After nearly two decades of fruitful interactions with the international community and additional programs at INTELSAT, I got an opportunity to participate in establishing a pioneering international system, WorldSpace, the world's first satellite digital radio system. This brought me in close contact with a different set of professionals in a startup enterprise as well as with professionals in many parts of Europe, Africa, and Asia, and it directly exposed me to the many unique characteristics of consumer-based systems. This program also led to a major role in the initial development and implementation of XM Radio, another digital radio system, this time in the United States. Both of these system-development efforts also enabled me to appreciate firsthand and absorb the merits of a close coupling between vision, mission, and strategy development on one hand and actual project implementations on the other. These systems, along with other contemporary digital radio systems, are presented as case studies in Chapter 10.

The final piece in my professional background relevant to this book was a pleasant lunch in early 2001 with an old colleague and friend from my INTELSAT days, Jeremy Allnutt, who invited me to take a graduate class at George Mason University, Virginia, and share some of my professional experiences with the students. After some hesitation, I accepted this invitation, and I am glad I did, since it provided an excellent opportunity to bring some coherence to the multitude of ideas and experiences one keeps collecting during a long professional career. It has also forced me to brush up on my own fundamentals, which in normal professional activities one often takes for granted.

The bulk of this book is devoted to fairly in-depth but largely managerial-level discussions around each of the major building blocks in an interactive process called the integrated business strategy process (IBSP). This process is first introduced in Chapter 3. As a prelude to this process, the first two chapters provide a quick background to the reader on the origin and brief history of satellite technology, its successes, and its trials and tribulations during over half a century.

After a very quick history, Chapter 1 introduces some of the predominant modes for satellite-based businesses. Using industry data as of the end of 2003 for various sectors of the industry, the reader is reminded of the importance of proper definition of what is accounted for and what is not in various published financial figures for the industry. The chapter ends with an introduction to the growing importance of direct-to-user (DTU) services and of matching changes necessary in management outlook and approaches to ensuring success.

Chapter 2 starts with a quick historical summary of the evolution of the various DTU services before moving on to the development of a few key lessons from recent industry projects and enterprises. Each of the eight lessons is followed by its own brief rationale. Almost all of the lessons are directed toward maximizing chances of success in future enterprises by learning meaningfully from the efforts in the past. This list of lessons does not claim to be either completely time invariant or absolute, nor is it meant to in any way cast a negative light on those who ventured into new areas to take this industry where in fact it belongs in the future.

Chapter 3 introduces the IBSP and provides a quick synopsis of the following eight chapters. A very short listing of the subject matter of these chapters is provided here as well.

Chapter 4 is devoted to the topic of strategy development. The standard definitions of such processes are recalled and illustrated with examples relevant to satellite-based systems. A couple of well-known satellite-based projects are presented as case studies for possible class or staff deliberations.

Chapter 5 delineates the importance of developing the business plan as a meaningful group effort, keeping in mind the specific objectives of such documents.

Chapter 6 is devoted to system planning, a function often with multiple hats and a binding force for the complete process. This chapter highlights some system optimization tools and recalls some interesting case studies.

Chapters 7 and 8 are devoted to the actual implementation of the system and its infrastructure. The emphasis here is more on managing complex programs to realize what is needed and not what may be technologically possible. The success criteria for a typical satellite-based project are also introduced here.

Chapter 9 approaches the whole process from the objective of managing for success. The lessons of Chapter 2 and the success criteria of Chapter 7 are revisited in the context of the IBSP. Attention is also drawn to traditional as well as not-so-well-known management tools.

Chapter 10 is a reasonably up-to-date review of the new and upcoming satellite-based service sector, digital radio. The readers will find this a good introduction to such businesses and hopefully will make their own analyses in the context of the processes described in this book.

Chapter 11 is on future evolution attempts to look ahead, drawing from the lessons of the recent past. It highlights recent technology and service projections by several experts and provides a general linkage into the future with the material presented in this book.

Acknowledgments

It is my pleasant duty to recognize and thank a fairly large number of professional colleagues, friends, and organizations who have helped me in various ways—some through exemplary leadership from the top and others by enabling me to discharge my responsibilities through outstanding contributions, innovations, and leaderships in their own domains. Still others helped during the process of writing this book itself.

Vikram Sarabhai, one of the most charismatic visionaries of the last century, was the founder of the Indian Space Research and was the one to push me into this fascinating technology in the late 1960s. Burt Edelson, the outstanding director of COMSAT Laboratories, made my awkward assignment of taking over the responsibilities for INTELSAT technology management a friendly one. Kishore Chitre, John Hampton, Pierre Madon, Gary Parson, and Emric Podracsky were instrumental in many ways in expanding the scope of my responsibilities at various stages. Noah Samara, another legendary visionary, provided many opportunities to play a very satisfying role in the creation of the new field of satellite radio systems. Of these leaders, it is sad to note that Burt Edelson, Emric Podracsky, and Vikram Sarabhai are no longer with us.

During recent years, I have also benefited a lot through my ongoing professional associations with John Dealy. He and his premier consulting team are a wonderful group with which to work and to learn firsthand the many ways strategy and engineering processes can interact to the benefit of the overall business.

Success as a manager hinges to a very large extent on your team. I have been fortunate to have the opportunity to lead a fairly large number of outstanding managers, many of them active industry leaders today. Those who contributed substantially to many of the concepts underscored in this book include Len Dest, Prakash Nadkarni, Pierre Neyret, Pat Rivalan, and Jeff Snyder.

Finally, I would be terribly remiss if did not acknowledge the many occasions when I was fortunate to benefit from the evergreen innovative talent and knowledge of the one and only Joe Campanella.

This book owes its origin to a suggestion from Bruce Elbert at our very first meeting a couple of years ago. Bruce, a reputed author himself of several excellent books, has assisted me at several stages, often completely in the background. I would hasten to acknowledge the equally valuable guidance (including patience with the missed deadlines!) from Barbara Lovenvirth, assistant editor, and Rebecca Allendorf, senior production editor at Artech House Publishers. They were always there to remind me what needed to be done, invariably with very helpful and pertinent suggestions. So was Mark Walsh, senior editor, through his periodic guidance and encouragement.

During the actual writing of the book, very often I would look for new data and information, and invariably many old friends and colleagues wholeheartedly came forward to assist. In this process, I also made several new friends. I would start with outstanding support provided by Olivier De Weck of the Massachusetts Institute of Technology, who assisted at several stages, including granting permission to include substantial extracts from his several publications and lecture notes. Several others were always there to look for the data I needed, often going out of their ways to be responsive. These include Jeremy Allnutt, Olivier Badard, David Bross, Olivier Courseille, Paul Dykewicz, Albert Heuberger, Bruce Jacobs, Tim Logue, Phil McLaster, and Andrea Maleter.

Many organizations and publishers were kind enough to give permissions to republish certain materials or adaptations thereof. In most cases, such permissions are acknowledged at the appropriate places in the book. These include the AES Journal, AIAA, Satellite Industry Association (SIA), Pearson Education, IEEE, Communication Research Center (CRC) of Canada, Futron Corp., ITU, McGraw Hill, Hughes Network Systems (HNS), MSV, Olivier de Weck, and John Wiley & Sons.

Finally, as acknowledged at the front of the book, I wish to thank my wife, Usha, for her patience with considerably higher than normal levels of absent mindedness and sullenness on my part during the process of writing this book. We are both fortunate to have three wonderful children, who in many ways inspire us in our golden years to continue to strive to do something worthwhile professionally.

1

Satellite Business Today

On December 31, 1901, Guglielmo Marconi demonstrated a whole new concept of communicating over vast distances *without* wires by transmitting a message from Cornwall, England, to Newfoundland, Canada, a distance of 2,100 miles. Over the next 100 years, this new *wireless* technology expanded beyond expectations and even today continues to enter progressively deeper in our personal and professional lives. In this long period of almost continuous innovation, several new words entered our everyday lexicon, and some of them left due to their subsequent obsolescence. Almost permanent entries include radio and television. More recent additions include microwave, pagers, and of course the cell phone.

About 50 years after Marconi's historic wireless demonstration, another word started its habitat in our vocabulary: satellites. While the basic concept of planetary satellites is as old as perhaps our creation itself, the concept of *artificial* Earth satellites is a much more recent one and was the culmination of the ongoing quest for better and better techniques to bridge vast distances via reliable and high-quality communications. When Arthur C. Clarke in 1945 conceived of the principle of geostationary satellites, he was postulating a global radio network [1]. Today, satellites are in use for a whole range of applications beyond radio, including telecommunications, broadcasting, and environmental studies, as well as for critical scientific and military needs. All of this has been achieved through a symbiosis of a number of disciplines, notably progressively more powerful rockets to escape Earth's gravity, highly reliable electronic and mechanical components for the satellites, and above all persistent efforts by many to establish viable and pertinent business models, often across international boundaries. In short, artificial satellites are credible businesses in their own right practically all over the world and hopefully one day even beyond.

Satellites Business—What Exactly Is It and What Is Its Future?

In early 2001, in the midst of the then-ongoing telecommunications and satellite industry downturns, a young engineer asked me, "Should I pursue a career in telecommunications or satellites?" My first reaction was to pick one of the two, but then after a few moments I opined, "You are comparing things that are not strictly equivalent. Telecommunications is a set of services, while satellites are *tools* for providing services. Services tend to have a much longer marketplace lifetime than tools; however, satellites have been pretty much resilient for almost 50 years and have generally shown an excellent ability to adapt to the changing needs of new services." While I felt pretty good about this seemingly erudite formulation, the young man perhaps left no less confused than he was earlier!

Satellites are essentially enablers of new services, and sometimes they can also make existing services better by replacing another medium or tool. Systems and enterprises where satellites of one kind or another are a major component are generally included in the category of *satellite businesses*, although the overall costs and revenues may be accounted for somewhere else. Depending upon their role, the end user may not even know the existence of satellites somewhere in the chain, unless of course the service provider has to explain away a sudden failure, something to which network television executives often resort. In other cases, satellites can be a prominent part of marketing the end-user services, such as direct broadcast of television and radio.

Implicit in the question from the young man was a broader one as well, triggered no doubt by some recent events within this industry—is there a future for satellites? The short answer is "yes," and we will devote a good part of this book to substantiate this assertion and suggest ways of ensuring it. At this juncture, we will underscore just two points: First, the unique and powerful attributes of ubiquity and multiple access that satellites alone can provide over vast areas will continue to play a key role in the longevity of this unique tool. Second, as long as the satellite technology continues to rise up to meet the ever-changing needs of the services *and* the related ventures are run primarily as businesses rather than mainly technology vindications, there will be a critical mass to the satellite business sector for a long time to come.

Satellite Business Boundaries

Compared to telecommunications and broadcasting, the aggregate of all satellite-based businesses is relatively small in terms of investments and revenues. Furthermore, because satellites are often part of a long end-to-end chain involving many entities, sometimes in different geographical areas, it can be difficult to figure out which part of the overall revenues is being counted as satellite business and which part is being left out. A good way to appreciate this often-overlooked aspect is to look at a few typical satellite-based businesses.

Figure 1.1 shows a few representative two-way services provided by satellites. Figure 1.1(a) is the classic preprivatization INTELSAT model. This international consortium owned only its space segment, and its revenues were limited to annual charges for the portions of the satellite resources leased to its participant owners or signatories [2]. Therefore, the INTELSAT revenue figures during that period did not include the much larger revenues derived by the member nations, or signatories, from the end-to-end telecommunication services provided to the users via the leased satellite capacity. Estimates for the INTELSAT revenues as a percentage of the corresponding end-to-end telecommunication service revenues varied from 1% to 7%. This multiplier factor over the cost of the satellite component underscores two points: it is extremely important to define what the revenue figures really represent, and satellites are often a financially small but critical enabler of certain services.

Figure 1.1(b) shows a typical corporate VSAT network, say for a retail store chain spread over a large geographical area. In this case all the costs, including satellite capacity lease, and revenues can be part of one business sector.

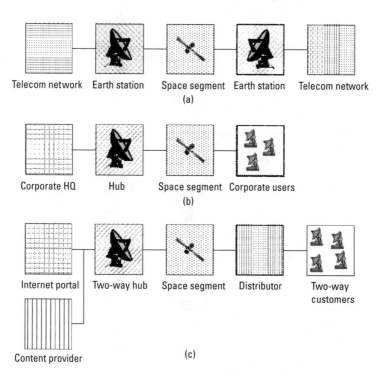

Telecom network　Earth station　Space segment　Earth station　Telecom network

(a)

Corporate HQ　Hub　Space segment　Corporate users

(b)

Internet portal　Two-way hub　Space segment　Distributor　Two-way customers

Content provider　(c)

Figure 1.1 Two-way systems—typical models: (a) classic INTELSAT model, (b) corporate VSAT network, and (c) two-way Internet service.

Figure 1.1(c) illustrates a business sector that is still evolving, namely broadband Internet access via satellites. One company can own all segments of the business, or alternatively it can just manage the space segment and leave other portions to different organizations. Obviously, the choices made will impact the revenues accountable as satellite business.

Figure 1.2 presents three somewhat analogous scenarios for satellite-based broadcast systems. Scenario (a) applies, for example, to the two major satellite television broadcasters in the United States, DirecTV and EchoStar. Both provide not only the broadcast function, but also the content, which is obtained in packaged form from specialized providers. While the receiver technology is proprietary to each system, they are manufactured under license and distributed by independent consumer-electronic companies. The new satellite radio companies, XM and Sirius, operate on a similar model, except at present they package a good part of the content in house, although this may change in the future.

Figure 1.2(b) shows a scenario wherein the satellite operator leases the capacity to multiple content providers who broadcast their programs to standardized home receivers. The services can be supported exclusively by advertising or as a mixture of advertisement revenues and subscription. Such models

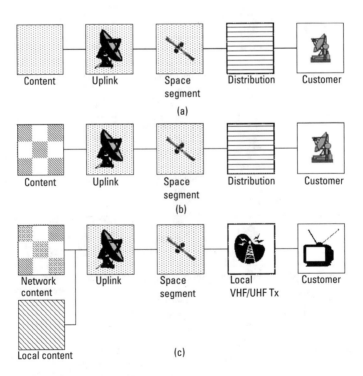

Figure 1.2 One-way systems—typical models: (a) semi-integrated broadcast system, (b) independent content providers, and (c) program distribution mode.

are quite common internationally. Figure 1.2(c) shows the scenario for television program distribution to local television transmitters. This is how satellites entered the television market, and this model is still in extensive use by networks in the United States and elsewhere and by many national and regional broadcasters.

These two sets of examples are only representatives of the different business models out there. They are presented to emphasize the importance of ensuring that equivalent revenues are being compared or added up for different entities or orbital locations [3].

Similar considerations apply when looking at the corresponding production figures and revenues on the industrial and manufacturing sides. It is not uncommon to find that some of the subcontractor revenues get counted twice. An extreme example, though not that rare, would be a prime contractor for a turnkey in-orbit satellite delivery counting the launch cost as part of his sale. At the same time, the launch agency may also include the same cost in its own sale figures.

Satellite Business Today

With these caveats, we now look at some of the key statistics for the *satellite business* up to 2003 made available by the Satellite Industry Association [4] and Futron Corporation [5]. Figure 1.3(a) shows the total annual real-year dollar figures on a worldwide basis for the industry over the period 1996 through 2003. As the bottom bar captures, the rate of change from 2002 to 2003 was the lowest over this period at less than 6%. In order to appreciate the significance of these numbers, it is helpful to look at the major categories behind these totals. Such a breakdown is shown in Figure 1.3(b) in four sectors: satellite services, satellite manufacturing, launch industry, and ground equipment manufacturing. It can be seen that the satellite services sector has continued to grow and in 2003 accounted for 60% of the revenues for the entire industry.

As we go further to the next level of detail, we get some additional insights—see Figure 1.3(c) for the satellite services sector. It can be seen that the subscription/retail services in 2003 are by far the dominant group, largely in the form of direct-to-user (DTU) television. While such DTU services are no doubt great success stories, the relatively large figures in Figure 1.3(c) are partly a result of the manner in which the revenues are accounted for different service groups. In order to better appreciate such factors, we now turn to Figure 1.3(d) presented recently by Futron Corporation [5]. This chart presents the utilization of transponders roughly over the same period as the earlier figures. Not surprisingly, the video sector—mainly distribution type—dominates the bandwidth

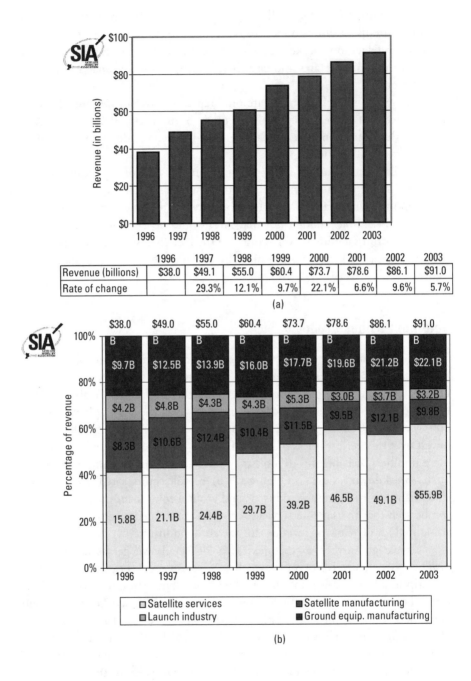

Figure 1.3 (a) Satellite industry revenues; (b) revenues by sector; (c) satellite service revenues; and (d) trends for transponders. [(a–c) Courtesy of Satellite Industry Association, (d) Courtesy of Futron Corporation.)]

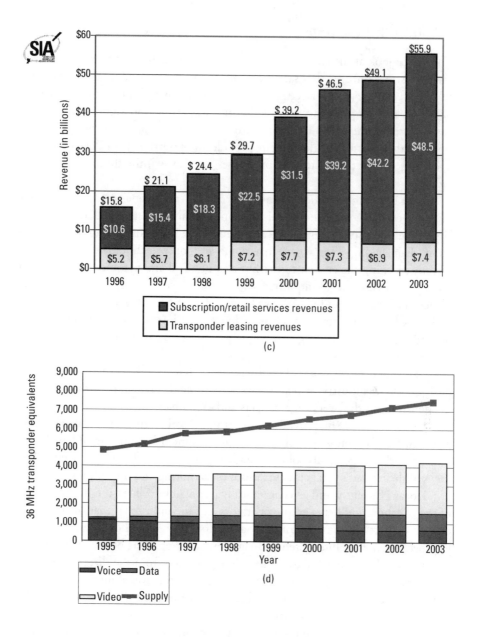

Figure 1.3 (Continued.)

utilization. If the *total* end-to-end revenues of all of the transponders were to be tallied up, the figures would be quite large and equally impressive.

Satellite Business Trends

This quick look at the changing dynamics of the industry is instructive in many ways. First, it confirms that satellites continue to be important, and often critical, enablers for several service groups. Second, bandwidth utilization by itself is not always a complete measure of the business scope and value of a satellite-based system. Third, DTU services are a great and hopefully growing success story for the industry, as they finally provide an opportunity for developing and operating complete end-to-end businesses entirely within the control of the industry.

The ongoing transition to complete end-to-end businesses is a direct result of half a century of technological development and experience derived from a large number of services in practically every geopolitical and economic environment. Painstaking efforts have overcome not only technological barriers to directly reaching the end customer but also political and commercial hurdles in order to provide such ubiquitous services in offices, homes, cars, boats, airplanes, and even to individuals walking and jogging. As is generally the case, such success in universality has been achieved not just through advances within the industry but also through an ongoing cooperation with consumer and content sectors.

This vital shift towards DTU services is changing the structure and composition of the industry in many ways. One notable change is in the nature of the demands on the program managers and executives.

A satellite system providing telecommunication trunking and television program distribution is typically built around standardized transponders that in today's world are generally sold and traded as commodities within a region. The focus of the related management teams is on acquiring and operating such transponders as efficiently and as economically as possible. A measure of the success of the collaboration between the operators and the industry is that the unit cost of such capacities has monotonically gone down for several decades. Nevertheless, because the satellite system managers are only enablers of the overall much larger value chain, they are to some extent protected from the direct scrutiny as well as vagaries of the commercial world. Finally, the availability of reasonably assured demand forecasts and fungible assets permits them to focus more on the satellites than on the ultimate users.

A DTU satellite business, on the other hand, is truly a self-contained end-to-end business, be it for television, radio, Internet, or any other such service in the future. As we saw in Figures 1.1 and 1.2, the business can literally encompass everything from live content generation by famous artists to a blown fuse in a hamlet in the middle of nowhere. Often the key driving factor is what the final user wants and what he or she is willing to pay for the services. While the satellite and other backroom technologies are no doubt key ingredients of the business, they are not the final determinants of success—rather, it is the user.

In summary, a DTU business manager not only has to cover a much wider range of disciplines and much larger financial amounts at risk, he has to be more right than wrong about what the final customer needs. In other words, the management team can no longer afford to be just *satellite centric*, it has to be *customer driven*. This theme underscores much of what follows in this book.

References

[1]   Clarke, A. C., "Extra Terrestrial Relays," *Wireless World,* October 1945, pp. 305–308.

[2]   Sachdev, D. K., "Historical Overview of the Intelsat System," *Journal of the Interplanetary Society,* Vol. 43, 1990, pp. 331–338.

[3]   Foley, T., "The World's Hottest Real Estate: Orbital Slots Are Prime Property," *Via Satellite,* Vol. XVI, No. 9, September 2001, p. 26.

[4]   http://www.www.sia.org/industry_overview.

[5]   McAlister, P., "Satellite Statistics: Is Recovery a Mirage?" *Satellite 2004 Conference,* Washington, D.C., March 2004.

2

Satellite Business Experience and Lessons

In the opening chapter, we took a top-level look at the status of the satellite business sector during the 2002–2003 time frame. In this chapter we will delve a bit deeper into some of these businesses, their evolution, and history, with a view toward picking a few key lessons going forward. These lessons in turn lead us in Chapter 3 to develop an integrated process that more closely couples different activities in a typical venture.

At the outset, it is pertinent to highlight that while looking at some of the recent pioneering systems, the text may sometimes appear critical in hindsight. This should in no way take away the industry's and the author's personal admiration for those organizations and for the leaders that took justifiable risks to venture into newer territories with a view to guide this technology into more competitive postures in the long run.

This chapter is in part an expansion and update of some of the material presented earlier [1].

Industry Evolution

Satellite technology applications started in the early 1960s less as business ventures but more as means of bridging many gaps in the ability to communicate over long distances, especially when separated by oceans or inhospitable land masses. Until then, more than half of the world's population in the developing countries depended almost exclusively on unreliable high-frequency (HF) links with limited reach and capacity that were subject to the propagation vagaries of

11

the lower atmosphere. While wideband multichannel microwave links were coming into use for domestic TV and telecommunication links, there was until then no available microwave tower tall enough to bridge the oceans or leap frog over undeveloped regions.

Against this background, the successful trials with the very first communication satellites [2, 3] were very quickly embraced by the telecommunication companies and nations around the world as a means of providing stable, high-quality long-distance circuits for communication and then for live relays of television programs. Because the early satellites were able to generate only very low power levels of radiated radio frequency (RF) signals, large antennas had to be installed on the ground to recoup the transmitted information from the extremely weak signal from the satellite. Even though the link capacities were still quite small, their strategic value to countries at both ends was quite large. As these links were much more stable and reliable than the old HF links, they were also quite lucrative and the telecommunication companies were happy to continue to operate them as integral parts of their business.

The rapid growth in long-distance traffic around the world put pressure on researchers to develop satellites with greater capacities. This in turn stimulated technology development across several fields, including more powerful rockets capable of launching bigger satellites with higher available solar power, newer frequency bands, and higher frequency reuse through advances in antenna technology. These bigger satellites not only became switchboards in the sky but also almost continually lowered the unit cost of service, thus further accelerating their overall progress in terms of telecommunication capabilities and revenues.

By the late 1970s and 1980s, the satellite technology had progressed to a point that serious consideration began to be given to DTU services, though mostly by independent technology development agencies and new entrepreneurs outside the "founder" fraternity of telecommunication companies and large national broadcasting organizations. While this was initially resisted by the incumbent operators of background trunking and television relay services, it was also clearly evident to system designers and business operators all over the world that satellite technology would be at its best for point-to-multipoint services, and a potential means of harnessing this unique capability was through DTU services.

Why were DTU systems attractive in the first place? For the technologists and system designers, the fundamental attraction was their ability to dramatically cut down the number of links from the source to the ultimate user. While locally broadcast and regionally relayed radio and television were already reaching the end users, it was the satellite medium alone that finally had the potential to directly reach literally millions of users over vast areas with truly high quality and greater choice of programming. For the business entrepreneurs, the DTU systems finally provided a means of controlling (and exploiting) the complete value chain of content, infrastructure, and advertising—all under one umbrella.

Satellite DTU businesses are currently evolving in four broad categories: television, mobile, digital radio, and broadband. The following paragraphs trace a short historical summary for each.

DTU Television Systems

For a considerable period after its invention, television was essentially local and simultaneous live broadcasts over wide areas were not technically feasible. Nationwide television distribution was first achieved through terrestrial microwave systems in the 1950s. However, such links required repeaters every 50 to 100 km and thus could not cross oceans or undeveloped land masses. This was one of the major incentives for the evolution of satellites, and in fact until this day the satellite payloads bear a strong architectural resemblance to their ancestral microwave links. (As a historical note, the microwave links themselves have practically lost the battle to optical fiber cables in many parts of the world.)

The satellite television distribution over major markets started via C-band, to be followed later by Ku-band, transmissions. Very soon, such broadcasts established their superiority over microwave links, as typically just one uplink transmission could reach all of the affiliated television stations through medium-size, often roof-mounted, receive-only antennas. While these broadcasts were strictly for affiliated TV stations, individual home dwellers also started installing such dishes in their backyards, particularly in rural areas not well served by television broadcasts. Even encryption of such broadcasts did not deter such individual, and often unauthorized, DTU receptions.

As the technology advanced, particularly in terms of maximum downlink satellite power capability, serious consideration was initiated in many parts of the world for direct broadcast to individual homes. National and regional development agencies funded qualification of critical items and flew some experimental satellites as well [4]. Proven broadcast technique at that time was analog, typically requiring 27 MHz of RF bandwidth for each television program channel. In order to reach users through small submeter antennas, extremely high satellite effective isotropic radiated power (EIRP) levels, about 65 dBW, had to be provided for each such transponder. These power levels were almost 10 to 20 dB higher than typical satellite transponders at that time. Even the most advanced spacecraft bus could accommodate only three to five television channels covering only a medium-size geographical area. One such early system was a joint German-French program TVSAT/TDF. The program ran into several technological and other problems, and after considerable delay two such satellites were eventually launched in the 1980s [5].

While such limited-capacity programs on both side of the Atlantic were solving their technical and other problems, the DTU television system (DTU-TV) planners had to resolve associated thorny political issues, primarily arising from apprehensions around the world of the "Big Brother" broadcasters reaching directly millions of homes across international boundaries. Furthermore, there was concern that developing countries would be left with no spectrum when they got around to investing in their own national broadcast systems. The solution finally adopted led to a *planned orbit,* where each nation was allocated a certain number of channels in the appropriate portion of the suitable geostationary orbit arc in a band exclusively set aside for satellite broadcasting [6].

Concurrent with this detailed spectrum planning, better receiver technology and some more innovative and less conservative link budgets began to substantially lower the technical barriers to DTU-TV, to the extent that medium-power traditional *fixed service* transponders could provide such a service with small dishes over reasonable-size coverage areas. This trend was pioneered by Astra in Europe and moved quickly elsewhere [7]. The channels were still analog, but a larger number could be provided with medium-power satellites than with the earlier "true" DTU satellites. This approach was so successful that additional satellites operating in different segments of the nominal 11-GHz band were collocated as needed to meet market demands. This approach continues into the future even today.

While such tens of analog satellite channels were adequate for several markets, they still could not compete against much larger number of channels provided by cable-television systems. Two almost simultaneous advances enabled realization of much higher capacities for satellite systems and at an affordable cost to the consumer. The first was digital compression that now enabled several TV programs to be squeezed into a single transponder without too much impact on subjective reception quality. The second was the feasibility of application-specific integrated circuit (ASIC) chipsets for the home receivers. These developments finally made DTU-TV with several tens of channels a reality [8].

DTU-TV is still largely a developed-world service. On both sides of the Atlantic, there are competing DTU operators, major among them being Astra and Eutelsat in Europe and EchoStar and DirecTV in the United States. Other countries with growing DTU-TV include Japan and several in Latin America. Despite its impressive growth, however, DTU-TV has still not captured significant markets in developing countries. This is in part due to the old "Big Brother" factor that may not easily go away, particularly in the post–September 11 environment. A more clearly identifiable factor is the cost of the consumer equipment. While this is often given away in developed markets against long-term service contracts, their costs are still too high a barrier to entry in several developing markets, particularly when compared to cable television. These costs must be brought down substantially through low-cost production and a fresh look at the overall system design, starting with consumer needs in each and every specific market.

Mobile Satellite Services

What was said about the relative isolation of developing countries connected only by the unreliable HF radio links prior to the advent of satellites was even truer for merchant navy ships and other unescorted vehicles in high seas. It was not that rare for such ships to be lost forever in the high seas under severe weather conditions due to total lack of any communications. The first serious effort was made by COMSAT over the Navy satellite, Marisat, in the late 1970s. Its success soon culminated in the formation of Inmarsat in 1979, an international treaty organization similar to Intelsat but with limited focus on connectivity to and from ships in waters around the world. This system has been nurtured ever since, in step with the growing demand, including additional ground-based terminals. Steady growth without flashy and risky steps has been the hallmark of this system. Successive generations of Inmarsat satellites have provided greater capacity and smaller and smarter user terminals. The service base has also been extended to land masses, along with additional services beyond narrowband voice. The system has established a niche for professionals, particularly those often traveling far away from modern telecommunications systems, both over distant waters and land masses.

For a variety of organizational and other structural reasons, Inmarsat by itself has not so far entered the field of truly mobile communication. Rather, it spun off a company, ICO, in 1995 to enter that market. As has been well documented elsewhere, a number of other ventures also entered this market. The success, or lack thereof, unfortunately has also been well documented and analyzed from a number of perspectives [9].

One such system, Iridium, was truly revolutionary in a number of ways. It was announced in the late 1980s as a bold concept for a virtually stand-alone system to provide mobile communication from anywhere to anywhere around the globe. Using a constellation of 66 satellites in polar low Earth orbits, the system aimed to create significant improvements in telecommunications in all countries big and small [10]. Being a global system, the whole system had to be created in one step at a cost of over $5 billion, a figure quite high even by today's standards. After a lot of debate about frequency allocations and the risky technology involved, the system was internationally funded not as a treaty organization but instead through agreements between operating entities around the world. Its potential success was impressive enough to stimulate several other similar systems. The industry saw here a growth opportunity and proceeded to create a capability to produce and launch a large number of satellites within short periods. In sum, the upside potential of the Iridium system was foreseen to be significant, both in terms of solving global mobility gaps and by acting as a catalyst for an industry somewhat numbed at that juncture by the prospect of the then-new optical fiber cables threatening to make satellite technology obsolete and redundant.

Unfortunately, the results were quite different and disappointing, not only for Iridium but also for the industry as a whole. The system suffered substantial programmatic delays and was finally commissioned in 1997. The user handsets were delayed and substantially bulky, especially when compared to the then fast-progressing cellular systems across cities around the world with smaller, elegant handsets with lower associated charges. The Iridium system finally could not be saved and entered bankruptcy in 2000 and is now operating on a limited scale for specialized services.

Digital Radio

Radio is what started it all, over a century ago, for the electronics, wireless, telecommunications, and entertainment industries. Yet in the DTU field, satellite radio was a latecomer compared to television.

In the context of this chapter, it is encouraging to see now that despite a late start and some tough business environments for some of the systems, satellite radio systems are operating in the United States and several parts of the world. This technology has a huge potential, given the large number of radio listeners around the world in cars, homes, and outdoor environments. Chapter 10 provides a comprehensive summary and an overview of all of the current systems in this field.

Internet Via Satellites or Broadband

In chronological sense, satellite broadband as a concept predates several other DTU applications of the satellite medium. In the 1990 timeframe, roughly when other nongeostationary orbit satellite systems for mobile applications were being pursued, Teledesic announced a truly revolutionary broadband system with potentially hundreds of lower orbit Ka-band satellites. The system was, however, never built, and after over a decade of modifications and deliberations, the company in 2002 returned the frequency licenses and suspended all work on this system. While this could be considered a setback by technology aficionados, from the business perspective it was a wise move to pull back investment if it did not appear justified.

As is well known, the explosive growth on the Internet has given a whole new meaning to the definition and business potential of broadband. This has stimulated several initiatives, several of which have not yet fully matured or played out in the marketplace.

It is not always appreciated, or even conceded, that the Internet in many ways is heavily dependent on the so-called old-fashioned telecommunication networks. It is in fact a tribute to the inherent flexibility of the bulk of the telecommunication links and pathways that they did quite easily adapt to the packet-switched nature of Internet-related traffic. Recognizing this, the international satellites were the first to get into the broadband business segment through the extension of Internet portals to different parts of the world. In many situations, they also helped by "broadcasting to the edge" for frequently requested large Web pages. This sector of the satellite business has seen a modest growth and has compensated to a certain extent for the drop in traditional telecommunication needs.

The next group of entries has come through an extension of Ku-band DTU-TV systems to provide two-way links for broadband access to homes and businesses. Initially, they used telephone lines for the return links, but this was quickly followed by genuine two-way links via satellites. Recognizing the earlier entries of digital subscriber line (DSL) and cable modems in most markets, these satellite entries targeted only those market segments that were outside the reach of the terrestrial alternatives, somewhat akin to the early attempts by television broadcast systems several years earlier. Not surprisingly, the results were also similar in the sense that the offerings were not very popular and did not grow in any significant way, partly due to the nonoptimum system configurations leading to high consumer costs.

The next generation of broadband systems to be rolled out in 2004 and 2005 has recognized the limitations of the earlier attempts for this application. While the three major contenders, iPSTAR, Spaceway, and Wildblue, differ in their detailed configurations, they do share the common objectives of offering optimized solutions capable of competing across the board with terrestrial alternatives in their respective markets. They hope to achieve these objectives by lowering the unit costs of service via high-capacity satellites and customer-driven user equipment [11–13]. We will revert to these aspects later in the book to evaluate the pros and cons of such strategies.

Lessons Learned

This quick journey through nearly 50 years of the evolution of satellite technology and associated businesses demonstrates at least one clear trend: A greater proportion of satellite-based businesses are moving closer to the ultimate consumer through stand-alone enterprises. This change has been quite dramatic in many ways, and it is not surprising therefore that not all such stand-alone, direct-to-the-consumer projects have been successful as businesses. The timing of such failures has been particularly unfortunate, as several such setbacks happened just before or even overlapping the much larger dotcom and telecommunication meltdowns around the turn of the century. The latter failures had in many cases also a knock-on impact on the satellite field because many of the stable businesses in the satellite field had a significant number of such telecommunication enterprises as customers or owners or both.

We will be reverting to different facets of this transition throughout this book, as in many ways it underscores the book's main theme of developing the optimum strategies and processes for business success. As a first step, we develop in this chapter a few simple lessons from both the good and the unfortunate experiences of recent decades. It is reiterated again that the sole objective of developing such lessons is to enable us to improve the probability of success in the future rather than to shed any negative or critical light on several pioneering ventures through the advantage of 20/20 hindsight.

Table 2.1 lists the major lessons. Each of these is elaborated in turn in the following paragraphs. Some of these lessons have been presented earlier [1].

Lesson 1—If You Are in a Consumer Business, Start with the Consumer

This lesson sounds so obvious that it needs some justification for its very inclusion in the list, particularly at the very top. However, recent experiences suggest that such a common-sense approach does need to be reiterated.

In order to fully appreciate this, we need to recall the manner in which the satellite industry has evolved over the first 40 years or so. As summarized in the earlier sections of this chapter, the industry started with telecommunication

Table 2.1

Lessons Learned

Lesson 1	If you are in a consumer business, start with the consumer.
Lesson 2	Reduce the lead times as much as possible or your business plan will become somebody else's.
Lesson 3	Unless you have positive confirmation of the business over the satellites' lifetime, avoid making nonfungible satellites.
Lesson 4	A system capable of growing in a modular fashion in the space segment has greater ability for midcourse corrections.
Lesson 5	Underserved markets are often also financially untenable.
Lesson 6	A successful business model in one part of the world can fail elsewhere.
Lesson 7	System dynamics of a two-way system is quite different from that of a one-way system.
Lesson 8	In good times and bad times, boring is often good [14].

trunking and television distribution. The standardized module for both of these applications soon became a 36-MHz transponder. This standardization was of tremendous benefit to the expansion of satellite systems worldwide. Understandably, the principal focus of most satellite programs was on the spacecraft design and technology, with the central objective of achieving the largest number of such transponders on each satellite with progressively better performance. In other words, the evolution became *satellite centric.* Such an approach has and continues to work well for organizations in the business of selling or leasing transponders, particularly since the unit cost of such transponders continued to drop.

However, the same approach when applied to the DTU services can sometimes have negative consequences, often quite serious. The senior executives heading the organization tend to give higher importance and priority to the space segment, often relegating the consumer equipment to either junior staff or even opting to come back to it later after the spacecraft design has been progressed much further along. Such managements tend to be much more familiar with the spacecraft industry than with the consumer industry, particularly if the suitable consumer-equipment producers and distributors are the larger consumer conglomerates, often overseas. At first glance, such an approach can also appear logical because the bulk of the capital investment of *their own company* is in the space segment, notwithstanding the fact that the cumulative cost of all of the consumer equipment and services—albeit by other companies and retailers—in a successful DTU project could exceed the initial capital investment related to the space segment.

A subset of such satellite-centric management focus is the glamour attached to the actual launch of the first satellite. Even after 50 years of satellite launches, a rocket launch to take a spacecraft to its orbital position is often billed or at least interpreted as the start of business itself, thus raising premature expectations. Unless the consumer units are already out there on retailers' shelves, all that actually happens for new ventures at first satellite launch is the start of the depreciation of the related investment, often without the start of revenues.

A quick look at some of the recent consumer-based projects would confirm where these factors had a dominant impact. Of course, the ultimate success or lack thereof is quite often a complex result of the impact of several factors and other lessons, some of which follow.

So what is a better approach, businesswise? First, progress the user equipment in accordance with market surveys for the specific service up to point where its cost, features, and market acceptability can be assessed with some accuracy. Conduct a detailed parametric analysis for the total system cost on the basis of the latest market dynamics, user equipment, operations, and total life cycle costs. Optimize the satellite design to match the best business scenario with acceptable technical and financial risk. While the satellite manufacturer(s) should be involved in this optimization process, the final contract should be signed only after this optimization process has been completed. Chapter 7 on engineering the system further elaborates these aspects.

Finally, paraphrasing the legendary Louis V. Gerstner, build the satellites from the customer back and not from the satellite out.

Lesson 2—Reduce the Lead Times as Much as Possible or Your Business Plan Will Become Somebody Else's

This lesson is not quite unique to satellite systems but applies to all businesses that are based on fast-developing innovation and technology. Once a project is funded, a common mistake is to forget that market projections are time dependent, particularly if alternative technologies are also entering the same market segment or existing technologies achieve a significant improvement in performance or cost, while the new venture is still going through its growing pains and tribulations.

This has happened quite often in the satellite industry, even before the advent of consumer-based systems. For example, in hindsight, the international satellite systems were slow to recognize in the early 1980s the likely impact of optical fiber technology for undersea cables. As a result, several of the options to purchase additional spacecraft in fairly expensive programs with relatively large nonrecurring costs could not be exercised, thus forcing these costs to be recovered from a smaller number of spacecraft.

More recently, the outcome of several mobile DTU satellite-based ventures has been well documented [9]. They represent a group where the collective impact of several complex factors had an unfortunate outcome. One such factor was the inordinate delay in bringing the projects to completion. At the time when some of these projects were initiated in the late 1980s and early 1990s, they had a real potential to establish a new paradigm of ubiquitous service worldwide, leapfrogging practically everything else that existed at that time. Unfortunately, by the time these systems came on stream, they not only had to bear the burden of Lesson 1, they also to face the harsh reality that the potential market had been almost usurped by the much faster pace of progress for terrestrial cellular systems.

Lesson 3—Unless You Have Positive Confirmation of the Business Potential over the Satellites' Lifetime, Avoid Making Nonfungible Satellites

Since fungibility is more of a financial term, it is perhaps appropriate to say a word or two about it upfront. Just like any other asset, a satellite system is considered to have a high degree of fungibility if it can be used for multiple applications, including some not even envisioned at the time of its design. Such a system carries a lower risk of ending up as a substantial financial loss. Conversely, a highly specialized satellite system optimized for just one application, often also with a specific set of modulation and access techniques, would be considered nonfungible and could entail high financial risks if there are uncertainties in the relevant demand materializing or migrating to other systems or media.

Almost right from their start in the early 1960s, most of the operational telecommunication satellite systems have achieved impressive fill factors. Even with largely committed usages, there was the inevitable uncertainty of traffic materializing close to the original planning and marketing assumptions. However, notwithstanding such uncertainties, two self-imposed disciplines contributed toward respectable fill factors and hence financial viability of successive generations. The first was the use of standardized 36-MHz transparent or bent-pipe transponders, mentioned earlier, which were largely interchangeable for a variety of services and applications. Even where larger bandwidth transponders were required for specific applications or bands, they were designed to approximate multiples of 36-MHz transponders in terms of their functionality. This feature allowed the operational teams to easily accommodate new patterns of traffic during the lifetimes of the satellites.

The second was the use of wide area coverages for most applications. While this often led to larger ground antennas than would have been required with more focused or spot beams, it provides excellent insurance against traffic vagaries and variability of the points of origin and termination of traffic streams. Thus, even today, hemispheric beams for international systems and Continental United States (CONUS) and other national or regional beams are a standard. A good conservative principle seems to be that you should not constrain the location and needs of a future customer.

As the digital technology emerged, innovative processing techniques, such as compression and digital speech interpolation, were introduced to multiply the throughput of the transponders designed long ago with only analog signals in mind. Inevitably, the benefits of processing onboard the satellite were soon identified as well. These included a dramatic reduction of uplink noise and distortion through onboard regeneration, dynamic switching of traffic streams between stations, and adoption of different multiplexing and multiple access techniques for the downlink and the uplink—all potentially adding up to higher capacities or lower ground costs or both.

Despite these potential benefits of onboard processing, most commercial systems were wary of adding such technologies onboard due to concerns that failures in such units could lead to substantial drop in overall usability of the spacecraft. There were also concerns that such units would inhibit the use of more advanced access and modulation techniques in the future. Accordingly, most of the initial applications were limited to technology and perhaps tactical missions. As the reliability of onboard units on such missions was demonstrated, several commercial satellites began to introduce onboard processing with some kind of fall-back arrangements for minimizing the impact of failures in these units. These included satellite-switched time-division multiple access (SS-TDMA) onboard the INTELSAT VI series of satellites and different types of processing units on several other programs.

The initial DTU systems, predominantly for television broadcasting, also adopted transparent architecture for the payload. However, due to the nature of the broadcasting function, the satellites could not be easily used for other applications in case the markets did not materialize. In other words, the satellites, despite adopting transparent-repeater architecture, were only partially fungible in financial terms.

Does that mean that satellites should forever remain of the vanilla transparent variety? Not necessarily, but the incorporation of any unique features onboard should be a conscious business decision with appropriate fallback positions and should not be dictated only by the attractiveness of new technology. However, once a market has been reasonably established, the second generation systems can introduce some elements of nonfungibility if it increases the business potential and reach. An example is the introduction of spot beams on second (or third) generation DTU-TV systems in the United States.

Some of the modern mobile systems are able to achieve their full potential only through substantial onboard processing. These include the Iridium, Inmarsat, and Thuraya systems, to name a few. By the very nature of such satellites and the associated spectrum allocations, some of kind of call or message switching onboard is absolutely necessary. The traditional approach is to place all of the processing onboard. However, recently serious consideration is being given to placing this switching on the ground. Apart from improving the fungibility of the satellites, it has the added advantages of releasing additional power in the satellite for more powerful downlink beams and allowing much more sophisticated processing and beam forming on the ground. However, it does come with the need for substantially higher feeder-link bandwidth.

In summary, there is a technical and business tradeoff between performance and fungibility, and in the system design the dominant guiding principle should be the confidence level in the business forecasts.

Lesson 4—A System Capable of Growing in a Modular Fashion in the Space Segment Has Greater Ability for Midcourse Corrections

For the last almost 50 years, it has been generally axiomatic to build the largest satellite with the largest number of transponders you can squeeze in. Given finite spectrum allocations at specific orbital locations, this has often required concurrent advances in onboard frequency reuse technology. Apart from the inevitable industry pressures to go for bigger spacecraft, the principal driver behind this trend was that for the still-significant transponder-leasing business, larger and larger spacecraft also generally led to a continuous drop in unit cost of transponders, often in absolute dollars and ignoring inflation. The alternative of achieving the same capacity through colocated satellites has generally not been in favor, primarily due to the high costs involved in duplicating complex antennas and other common subsystems on every satellite in the cluster.

For the DTU business, however, particularly for the ones breaking new grounds, this approach can be financial risky. Building the largest satellite and filling up all of the spectrum allocation also means sinking almost all of the capital investment upfront with its concomitant larger risk if the market does not materialize as anticipated.

The DTU-TV business illustrates this principal through several successful examples around the world. The most successful is also the oldest. The Astra system built the in-orbit capacity in a modular fashion in step with the increase in market demand. Instead of one large satellite at one orbital location, it chose to build a cluster of several small satellites launched staggering over several years [7]. While this approach was initially dictated by the limitations of spectrum availability, the operator has maintained this approach even today, as it not only spreads the investment over a longer period, it also permits progressive introduction of newer technologies as they mature. Several other systems around the world have emulated this approach of balanced, financially friendly system growth.

In sum, wherever feasible, adopt a modular approach in consonance with market growth. The regulators can assist in this process by not insisting that the entire allocated spectrum should be used at the very start.

Lesson 5—Underserved Market Segments Are Often Also Financially Untenable

Quite early in its evolution, the satellite medium demonstrated the ability to reach the far corners of the Earth that were within its coverage areas. This unique capability has often encouraged competitive media and even well-known industry and regulatory prognosticators to sometimes relegate the satellite medium to the so-called *thin* routes, thus leaving the far more lucrative businesses to alternative media—generally different types of cables. While the latter do and can achieve very high capacities, it is not correct to underestimate what the satellite medium can also do in such applications as well.

Some of these misconceptions about the satellite medium have in fact been unintentionally perpetuated by the industry itself by venturing into certain new fields prematurely. The classic example we have already touched upon earlier in this chapter is the direct broadcast of television programs. By attempting to provide such services with analog techniques and limited spectrum, the early systems came up with offerings that could not be viable businesses on a stand-alone basis. It was only when the trio of technological advances—power, signal compression, and ASICs—were harnessed through a sound system design, that the satellite medium could not only compete aggressively in direct broadcast of television programs but also often beat in quality and choice the cable alternatives, which were slow to adopt digital technologies.

As often happens, some lessons are to be learned twice, and we did start to repeat history in the broadband arena. The early attempts to provide broadband by leveraging the DTU-TV Ku-band satellites were only modestly successful, and yet in almost every forum there was a lot of advice given to stay in those market segments that neither the cables nor DSL coveted; in other words, stay in what could turn out to be untenable markets. Fortunately, several industry leaders have recognized this and are urging the industry to come forward with solutions that will compete head on with alternative media in their backyards as well. The years 2004 and 2005 will see at least three such attempts through the IPSTAR, Spaceway, and the Wildblue programs [11–13]. While they do not fully comply with some of the other lessons we are postulating here, they are indeed targeting markets that are potentially tenable.

Lesson 6—A Successful Business Model in One Part of the World Can Fail Elsewhere

Satellite systems have traditionally achieved impressive manufacturing learning-curve benefits in costs through a series of identical or nearly identical satellites. This has been the practice with international systems in particular. In such organizations, however, the overall accounting tends to blur the fact that some of the satellites in the series do not really match the market conditions in different parts of the world.

However, for systems planned for direct consumer services, the economic risks are far greater than those for trunking services. For example, the market sensitivity to a common consumer equipment price across the globe can be very diverse from one region to another. Therefore, while *the link budgets are the same everywhere, the business models may not be*. What looks cash flow positive in an affluent part of the world can be a financial liability in another continent. Therefore, while there could be tangible synergy and economic benefits in developing a common series of satellites, the applicable business models can be very regional, if not local. These should be addressed upfront and weighed against the initial development costs.

Lesson 7—System Dynamics of a Two-Way System Is Quite Different from That of a One-Way System

A powerful and unique attribute of a satellite broadcast system is that the system infrastructure cost is independent of the size of the audience. Once the system has been implemented, the ongoing costs are only limited to those associated with the acquisition of customers. This is one of the reasons that satellite-based television systems have a distinct edge over cable systems.

When the demand for DTU Internet service began to grow, it was at first glance quite logical to offer such services as an extension of television services. As long as the return channel from the users was the telephone line, this was indeed the right approach. However, the cost of the phone line and the associated Internet service providers (ISPs) made such services noncompetitive with other alternatives such as DSL and cable modems.

In order to make the services more attractive, the concept of using satellites in both directions is the current approach. However, now the system is no longer pure broadcast, as the satellite capacity needed to accommodate the uplink return signals from each user is now a function of the market size. In fact, for Ku-band applications, these capacity needs soon started dictating the overall economics, partly due to the relatively inefficient transmissions from the small dishes at users' premises.

While planning two-way services as a business, therefore, it is important to keep in mind that unlike broadcast systems, the space segment costs will increase as the market size grows.

Lesson 8—In Good Times and Bad Times, Boring Is Often Good

The last lesson is in fact a quote from a recent speech by a senior executive at a satellite conference [14]. The comment was made in the context of a rather dismal picture in 2002 for both the telecommunication and satellite industries, but it can have applicability under other circumstances as well.

As this chapter has tried to capture, the unfortunate outcome of several recent satellite-based ventures has been often due to a rather aggressive approach towards technology, mostly of the satellite-centric variety. Often the entries into new risky ventures have been made without adequate and dispassionate analysis of business prospects. As the industry has transitioned from trunking and distribution to DTU services, attention to the users' needs and service cost effectiveness has not always been the first priority. Often the dissatisfaction was noticed only after it was too late to remedy the matters.

The key factor that distinguishes the satellite medium from other technologies in overlapping market segments is that the bulk of the capital investment is upfront, and any mistakes in configuring the space segment generally cannot be corrected. At the same time, the market changes and surprises are inevitable in these fast-moving times. Therefore, a safe and financially prudent approach often is to design space segment with at least a minimum level of flexibility and some fungibility to earn other types of revenues if required. Such an approach can often lead to bland or boring designs. However, such cosmetic attributes are forgotten if the bottom line is business success.

In summary, the sophistication of the system should be only what the business dictates. If such a system appears "boring" to technophiles, but makes the investors happy with sound returns, so be it.

References

[1]	Sachdev, D. K., "Three Growth Engines for Satellite Systems," *AIAA 20th International Communications Satellite Systems Conference*, Montreal, Canada, May 12–15, 2002.

[2]	Davis, M. I., and G. N. Krassner, "SCORE—First Communication Satellite," *Journal of the American Rocket Society*, Vol. 4, May 1959.

[3]	Holahan, J., "Telstar, Toward Long-Term Communication Satellites," *Space/Aeronautics*, Vol. 37, No. 5, May 1962.

[4]	Sabelhaus, A. B., "Application Technology Satellites F and G," *Astronautics and Aeronautics*, Vol. 9, September 1971.

[5]	Kaltschmidt, H., "The German-French TV-Sat/TDF1 System," *IEEE Canadian Communications and Power Conference*, Montreal, Canada, October 1980.

[6]	World Administrative Radio Conference for Planning of the 11.7- to 12.5-GHz Band, ITU, Geneva, Switzerland, 1977.

[7]	"The Astra Satellite: Innovative Commercial Applications," *Space Communications and Broadcasting*, Vol. 6, No. 1–2, May 1988.

[8]	Lucachick, P. S., "The Technology of DBS in America," *AIAA 14th International Communications Satellite Systems Conference*, Oakland, CA, March 1992.

[9]	Richharia, M., *Mobile Satellite Communications*, Reading, MA: Addison-Wesley, 2001.

[10]	Sterling, D. E., and J. E. Hatlelid, "The Iridium System—A Revolutionary Satellite Communications System Developed with Innovative Applications of Technology," *Milcom '91 Conference Record*, paper 21.3, McLean, VA, November 1991.

[11]	Swakpan, T., "The IPSTAR Broadband Satellite Project," *21st AIAA International Communications Satellite Systems Conference*, paper 2003–2206, April 2003.

[12]	http://www.wildblue.com.

[13]	Sarraf, J., "The Spaceway System: A Service Providers' Perspective," *IEE Seminar on Broadband Satellite: The Critical Success Factors—Technology, Services, and Markets*, October 17, 2000, pp. 15/1–15/6.

[14]	Olmstead, D., Keynote Address at *International Satellite and Communication Expo (ISCe) Conference*, Long Beach, CA, August 2002.

3

Integrated Business Strategy Process

In the first two chapters, we have taken a quick survey of the evolution of the satellite industry and its services. We have also noted a clear transition underway toward DTU services. Not unlike any other field, over the years there have been successes and failures. From these we have derived some general lessons to take forward. The rest of the book will be devoted to the development of and means for maximizing the probability of success. This chapter sets the framework toward this objective.

Success or failure in a large program or project, involving several disciplines and spread over several years' duration, is often difficult to pin down to specific factors, groups, or disciplines, although there is a great temptation to do so. Thus, when some of the satellite systems that had ventured for the first time into the commercial use of low-earth orbits (LEO) had problems, there was an immediate banishment of such orbits from any further consideration. Along the same lines, when tragic accidents took place in the shuttle program, there was an instant cacophony of condemnation of the NASA culture and so on. Nevertheless, when we go beyond such snap judgments, we often do find some common and generic factors in addition to specific technologies or system architectures adopted.

Leadership and management skills at various levels can and do play important roles, and we will address their importance later in the book. For enterprises or projects involving fast-changing needs of the ultimate consumer, such as stand-alone DTU satellite systems, a mismatch between the changing market needs on one hand and the evolution of the infrastructure on the other can often become an important element in the final determination of success or lack thereof. Sometimes such mismatches can be a result of external factors beyond the control of the company; however, more often than not they are due to the *stow-pipe effect*, or inadequate *peer-to-peer* timely interactions between various functions. Experience in a wide range of circumstances suggests that, in most cases, deliberate structural mechanisms are necessary to ensure such interactions, and mere periodic edicts or directives from the top for cooperation are generally inadequate. It is worth noting that such "inadequacies" are by no means unique to individual industries, they are also periodically identified at all levels of society in different parts of the world.

Integrated Business Strategy Process

Most modern organizations with good leadership and management skills at the top generally have a clear vision of what they are or aspire to be. From this vision, they derive a well-thought-out set of objectives and associated strategies. Often such strategies require specific projects to be implemented for new or improved products and services. These projects are self-contained engineering-oriented programs with their own objectives and schedules. In a typical organization, the business and strategy team would present its results to the top management, proposing certain projects to be implemented by the engineering groups. While there would have been some discussions with the engineers about the ballpark costs, more often than not the business teams would have made their own assumptions in their strategy-development process. Along the same vein, once the engineers are allocated the project (with the associated expenditure authorizations of course), they also tend to run with it on their own without too much involvement of other functions.

During the period when a typical multiyear project is being implemented, at least two categories of changes may be afoot with little or no knowledge of each other's existence. The market could be changing for any number of reasons, but such changes may or may not be communicated directly to the project teams that are up and running. Along the same vein, the engineering teams often come across "fascinating" new technologies and adopt them by putting on their own "marketing hats." As long as they are within the overall budgets, no waves are created. Of course, both of these sets of changes in the market and in the system technical configuration do get reported up the stove pipes, but at the levels where both are read, the "dots are rarely connected," to borrow a popular phrase.

This scenario is not as far-fetched as it may seem at first glance. In fact, unless certain deliberate steps are taken, such scenarios can exist even when the different functions are located physically close to each other.

The remedy is in principle straightforward: create structures that force and nurture interactions. For this, we turn to standard treatments in a large body of

stem planning and other functions with regard to the capabilities and limita-
ns of the satellite medium. Appreciation of the regulatory environment and
tential early actions with regard to orbital locations and spectrum is another.
he market and competition analysis must also identify the appropriate time
indow for the project. Internal environment analysis should include an impar-
al assessment of the capabilities and limitations of the in-house team and its
dvisors. If key elements are missing, there should be a quick action to remedy
uch gaps. Intellectual property analysis is also an important component of
nternal environment analysis.

The winnowing process of alternative strategic options is another critical
tage where inputs from functions like system planning and engineering are
almost indispensable, especially for the estimation of the costs, schedules, and
capabilities of different options. The final process of selection of the business
strategy of the company has to be a true team effort in order to ensure that no
gaps exist in any of the business, system, and technical analyses.

The chapter concludes with a few examples of test criteria that a selected
business strategy for a satellite system should meet.

Chapter 5 on business plans addresses a topic that is rarely in an organizational process, but is instead handled by a select few executives presumably to
protect sensitive data from potential competitors. However, such an approach
can be counterproductive, both in terms of the quality of the plan itself as well as
in terms of inadequate appreciation of the overall company objectives by all of
the functional units. The correct approach is to involve the senior managers of
all units, with the proper instructions regarding confidentiality.

This chapter first draws attention to the often-ignored fact that a business
plan is not just for raising capital. It can also have other objectives, which can
influence the manner in which such plans are put together. Such objectives can
include attracting new partners to complement the expertise of the founder
team and motivating company personnel toward fulfillment of company
objectives.

Preparation of a business plan document should be a true team effort,
even more so than that for business strategy. Practically the entire organization
can and should provide input. This chapter also delineates what should go in
different sections of a good business plan document. In particular, attention is
drawn to the importance of a good executive summary. The emphasis throughout is on sufficient depth in all aspects of the document without too much
exaggeration or hand waving, either with regard to technology, deliverables, or
costs. The importance of full disclosure on all types of risks, not just launch-
related ones, is highlighted. In many ways, a good business plan document
can capture the good attributes of a management process like the IBSP and
can in fact be a microcosm of what the organization already is or wishes to
become.

excellent books for strategy and project management. Figure 3.
simplified version of a process typically used for the derivation
meet the company's mission. Along the same lines, Figure 3.1(b)
abridged sequence for project implementation. The first step
interactive process is to merge the two as shown in Figure 3.1(c).
looks simple on paper but can need a deliberate effort, sometime:
an organizational cultural change, to implement. In the next six
will go through this process, labeled *integrated business strategy*
one building block at a time, highlighting the interaction path
resultant benefits.

The following paragraphs now summarize the objectives and
the remaining chapters in the book.

Chapter 4 on business strategy development focuses on the s
of the IBSP, the development of a business strategy consistent with
of the enterprise *and* the environments in which it has to operate a
Proper development of the mission statement by itself is critical.
should be responsive at least to its customers, investors, and empl
external and internal environments must be carefully analyzed at this
components of the external environment include government fisc
potential markets, competitors, and applicable regulations. Developr
market share that can be captured is the first of many important s
business strategy development can and should benefit from intera

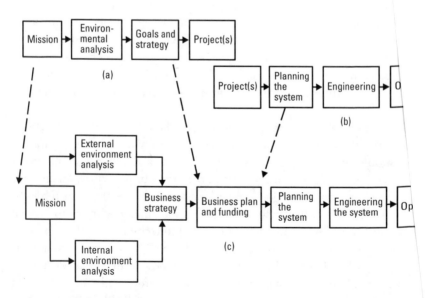

Figure 3.1 Evolution of IBSP: (a) corporate strategy process, (b) project process, and
IBSP.

Chapter 6 on system planning addresses an activity that can sometimes go by other names, including system engineering or project office. Such functions are invariably critical to the ultimate success but can often go underappreciated, as they operate mostly in the background. It is not surprising, therefore, that their importance often gets recognized belatedly by their absence!

In the context of the IBSP, system planning fulfills many functions through well-balanced expertise across practically all of the areas involved. For maximum effectiveness, such groups should be manned by experienced personnel with good interpersonal skills and familiarity with multiple jargons across the spectrum of disciplines. Through such expertise, this group can convert the marketing data to the engineering terminologies and thus assists the business strategy and business planning groups on alternative means of meeting the identified market needs and their costs, risks, and growth capabilities. It also has generally good inputs to make toward sensitive competitive analyses. Turning around its role and perhaps the jargon, it also works closely with engineering through the architecture design and implementation phases. One of the critical functions of system planning is the optimization of the overall system architecture; often, such exercises can not only provide a better insight to several system drivers but can add a competitive advantage to the venture.

The chapter provides case histories to illustrate the critical role of system planning and of a program that contributed to development of some of the concepts presented in this book. Another case history draws attention to some recent publications describing powerful system simulation tools and optimization models.

Chapter 7 on engineering the system presents in some detail the importance of a disciplined approach to the design and implementation of the system's infrastructure. At the outset, the importance of adhering to the principle of building what is needed and not what is technologically possible is highlighted. Key *measures of success* that are applicable to the whole IBSP are developed in this chapter, as this is where major irrevocable financial commitments get made.

The chapter then surveys top-level architectures for the major applications of satellite systems with a view to identifying the key common building blocks for such systems. Each of these blocks is then addressed in turn, focusing more on programmatic and management aspects rather than on detailed engineering.

Under consumer equipment, the chapter underscores the advantages of a customer-centric approach over satellite-centric design in terms of a better ability to meet customers' needs and in terms of minimizing major financial viability problems downstream. General guidelines are then developed for consumer equipment for different applications.

The same approach is extended to the discussion of spacecraft engineering. First, the system planning inputs and their impact on spacecraft design, complexity, and costs are summarized for different types of applications. This is followed by a top-level review of functions, relative complexity, and impact for the key subsystems, provided separately for typical payloads and platforms. Next is a general discussion and suggested guidelines for efficient spacecraft procurement and program oversight processes. The chapter concludes with a brief discussion of launch services and earth stations on similar lines.

Chapter 8 on system operations complements Chapter 7 and summarizes the key objectives for operational facilities. Depending on the type of service provided, the complexity and duties of system operations can vary substantially. The critical role of satellite control networks and their responsibilities and risks are delineated. Often the key determinants here are the extent of care taken in training personnel and efficient collection and accessibility of historical data on the managed spacecraft.

Increasingly, system operations include responsibilities for content in various forms. This can vary from the classical backhauling of programs from distribution satellites to the creation of custom content ensembles that can be a key factor for success in a competitive environment. Broadband services are rapidly evolving, with different companies having different demands from system operations. The higher importance of quality of service (QoS) for such services can be discerned from the plans for several systems to be deployed in the 2004–2005 time frames.

The chapter also underscores the substantially increased criticality of an efficient customer service for DTU services. Such functions also contribute to the IBSP through real-time feedbacks on QoS as well as with valuable customer inputs regarding follow-on capabilities.

Chapter 9 on managing for success takes stock of all of the preceding chapters and recalls key tangible and intangible factors that can contribute to ultimate success in an enterprise. On top of this list are leadership and management. The discussion highlights their commonalities as well as differences in such skills. Key managerial duties and responsibilities are recalled thereafter. Attention is also drawn to the fact that no manager functions in isolation, as all need what are popularly known as a *network of relationships* and *tradable currencies* with peers and groups within and outside the company.

The IBSP benefits as described in the preceding six chapters are then recognized. In particular, the IBSP chart in Figure 3.1(c) is redrawn, now including some of the major interaction paths between various groups and the resultant benefits in terms of maximizing the chances for success. From the many traditional tools for efficient program management, specific emphasis is placed on the usefulness of *earned-value* charts toward providing accurate real-time information on true progress as it happens. The chapter concludes with a recall of measures of success as discussed in Chapter 7 and the lessons derived in Chapter 2.

Chapter 10 on digital radio system case studies presents in some detail the evolution of this relatively new application of the satellite medium with a view toward illustrating some of the concepts and principles developed in this book. As the technology and the related systems are still evolving, this chapter runs the risk of being obsolete in some of its details.

In essence, satellite radio systems are a direct vindication of Arthur Clarke's vision 50 years earlier. The technologies and system concepts have evolved almost on a global basis, although the business models in different parts of the world are understandably different. A common factor, though, is almost universal appreciation of the quality, choice, and ubiquity of satellite-based systems as compared to the traditional terrestrial radio systems.

The chapter recognizes multiple visions pushing this technology, ranging from replacement of the venerable HF systems, entertainment, and news, to automobiles, and fulfillment of the terrible gaps in information availability in different parts of the world. The systems developed so far demonstrate the impact of each of these visions.

The systems described in this chapter include the pioneering efforts in Europe, World Space, Sirius Radio, XM Radio, and the MBSat System by MBCO of Japan and TU Media Corp. of Korea. It also includes a brief mention of the recently introduced digital techniques in traditional AM/FM bands.

Chapter 11 on future evolution revisits the very beginnings of the satellite technology and attempts to project what the future could bring.

The chapter considers in turn the three major contributors to future evolution: technology, services, and business strategy. In terms of technologies, the drivers appear to originate from broadband and to some extent mobile services. In order to meet the competitive challenges head on—rather than settle for potentially untenable markets—the focus is on reducing the unit cost of service. The suitable technological approaches identified for both of the areas are larger spacecraft with multiple beams and onboard processing.

In terms of services, there is a clear recognition that the center of gravity has shifted from the traditional telecommunications to broadcasting, broadband, and mobile. With increasing interest in *bundling*, it is quite likely that many ventures may provide two of these three or even all three segments.

The principles of business strategy, at their core at least, do not need to dramatically change. Obviously, application of some of the large spacecraft concepts, with extensive reuse and processing, underscores the increased importance of risk analyses. It also vindicated the importance of interaction among all functions through processes like the IBSP.

The chapter and the book end with a genuinely optimistic note, relying largely on the intrinsic strength of the medium and the demonstrated ability of the entire community for several decades to innovate continuously and change when so justified.

4

Business Strategy Development

Starting with this chapter, we will expand on the IBSP shown in Figure 3.1. Wherever applicable, we will draw upon the lessons developed in Chapter 2.

This chapter will address what is often the most difficult—and often short changed—part of the IBSP. An enthusiastic and genuinely fired-up team, full of entrepreneurial spirit, develops a concept or an approach and is extremely keen to start implementing it, lest the delay allow competition to enter the market first. Often, the founder team members may have left a safe-haven career in an established company and are in a hurry to vindicate their midlife risky decisions by quickly implementing their new project. In short, they are extremely keen to get out of their basement or garage office and make their dream company a reality!

So what is wrong with genuinely enthusiastic entrepreneurship? Nothing in principle; however, rushing to implementation (assuming there is funding, of course) can often be accompanied by a tendency to see only the positive external factors and to ignore the inevitable warning signs out there.

The portion of the IBSP that captures the critical activities prior to implementation starts with the mission and goes up to the development of business strategy. Figure 4.1 presents an expanded view of these elements, which are the subject of this chapter.

Mission

Mission is what the whole company or enterprise is about. Unfortunately, mission statements are often seen as some kind of window dressing for the top levels of an organization, to be pulled out of the drawer only when visitors come to

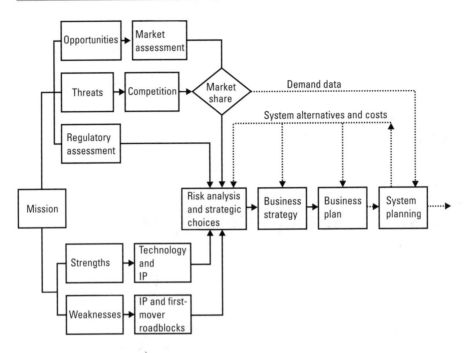

Figure 4.1 Strategy development process.

meet senior management and to be forgotten thereafter. Often, except for the core assistants close to the top management, most of the staff often does not even know what the company's current written mission statement is. This situation is often symptomatic of the lack of congruence between how the top management and the working level staff see where the organization is headed.

Even in well-run and tight-knit organizations, until recently, mission statements have deliberately been kept as general as possible, perhaps to give top management enough flexibility to make changes in the company's direction without having to coordinate every time with the different stakeholders in the company. However, this approach is gradually giving way to make the mission statement as clear and specific as possible. This is particularly true for new ventures directed at a specific market or designed to exploit a specific technology or intellectual property as a first entrant in the market.

According to Gray and Larson [1], "the mission identifies what we want to become or the *raison d'être*... a written mission statement provides focus for decision-making when shared with managers and employees. Everyone in the organization should be keenly aware of the mission."

Aaker [2] compares the need for a clear mission statement for a company with the types of questions we are all faced with in our childhood about our future, both in our own eyes as well as in the eyes of the society around us. He advises that "the mission should be conceptualized as a dynamic rather than a static construct… with a dynamic focus the mission will be a better vehicle to generate and screen strategies." In other words, a mission statement has to capture the essential reasons why a company exists or wishes to exist.

Wickham [3] has put together in one place a large number of business case histories that cover practically the whole canvass of situations that businesses in different industries can expect to encounter. Accompanying these case histories are clearly developed summaries of the basic underlying principles for strategy development and decision making. According to Wickham [3], a mission statement should be a "simple, easily remembered, impactful statement which defines the business's role in the world and what it wishes to achieve in the way of success."

A good mission statement has to clearly recognize and demonstrate who the company management is working for. As the age-old saying goes, "the man who pays the piper calls the tune"—in other words, as long as you keep the investors happy (whether private investors or the stock market at large), the common perception is that you are fully responsive. However, investors are not the only principal stakeholders in the broader sense of the company's mission. It is also equally important to recognize the company's customers as well as the employees as critical stakeholders. Wickham [3] adds to this list the company's suppliers and the government and wider community in which the company carries out its operations.

So what should a good mission statement include? Here the references quoted earlier have a large degree of commonality. Gray and Larson [1] identify its traditional components as major products and services, target customers and markets, and geographical domain; frequently, these could also include organizational philosophies, key technologies, public image, and contributions to society. Wickham [3, 4] emphasizes the scope of the product and its markets, how the company intends to compete, and the company's overall aspirations and values.

These lists might give the impression that the mission statement should be long and wordy. That need not be the case in almost all situations. What is important is that such a statement should be specific, not vague, and should clearly summarize the management's unambiguous sense of defined goals and objectives as well as its commitment to achieve them.

A remarkably short, yet quite responsive, example of a good mission statement was provided recently by David Weaver, a student in the author's class on project management for his class project [5]:

> To increase MobileMemory's market value in the Americas by providing our industrial and government customers with simple, secure, mobile data repositories produced by a team of dedicated and satisfied employees.

Weaver went on to explain [5] that the mission statement addresses the shareholders by increasing MobileMemory's market value. It addresses the customers by mentioning new markets. It recognizes the employees' dedication as well.

Environment Analysis

A good mission statement and its goals and objectives set the stage for developing strategic choices from which a sound business strategy would emerge. In order to arrive at the applicable strategic choices for a particular mission, a series of critical analyses need to be carried out, as shown in Figure 4.1. Following the mission, on the top are shown what were collectively called in Figure 3.1 *external environment factors*, while at the bottom are typical *internal environment factors*. A careful analysis of each of these factors, *interactively with other IBSP functions*, should generally lead to a balanced set of strategic choices.

External Environment Factors

The external factors shown in Figure 4.1 are what could be called *micro-environment* factors. In addition to these, there are other broader sets of *macro-environment* factors related to the business, political, and social environment in which the company has to operate. The latter can vary quite a bit in different parts of the world and in their impact on different industry segments. They can of course also change as a function of time (e.g., as a result of government policies). Such policies can often change in the middle of major long-term projects, causing complications of varying degrees. For satellite-based systems, the more important macro-environment factors can include interest and exchange rates, taxation and tariff polices [3], and all shades of restrictions on the release of product documents for national security reasons. Many of these factors can impact the actual cost of the project. In some cases, they can favorably or adversely impact the market being targeted. For example, changes in import tariffs can significantly change the cost of the DTU consumer equipment, which may in turn impact the market forecast itself.

We will now consider the other external factors shown in Figure 4.1.

Market Assessment

In each field, some market segments are relatively steady and established. In many cases, they follow commonly accepted standards, which can give them a flavor of a commodity business wherein price, availability, and QoS become the main distinguishing factors between various suppliers. In such steady fields, the market analysis is also relatively static and may not materially change over the duration of a project for a new entrant. This is true to some extent for the satellite transponder-lease business, although newer satellites can sometimes have a marketing edge via higher power and better performance.

At the other end of the spectrum are those market segments where either a brand new or at least relatively new service is planned to be offered. In a completely new market segment, it is not uncommon for each entrant to have his own proprietary equipment, partly because there is no industrywide standard yet. Once a new company has some success, it in fact resists any standardization because it might lose its marketing edge and wishes to avoid becoming a commodity. In general, the huge and fairly homogeneous market in the United States tends to favor such an approach, although the recent experiences with *wi-fi* wireless and the cable modem standards are convincing evidence of the long-term benefits of standardization through rapid expansion of the market size.

For a new entrepreneur, a proper and thorough market assessment is absolutely essential, preferably by a totally independent source in order to avoid any risk of the founding group even unwittingly biasing the results toward their own perceptions or desires. Such an assessment should try to define not only a possible market share but also its sensitivity to timeframes, particularly with regard to any concurrent developments underway in competing alternatives. It should also include estimates for the likely costs related to customer acquisition and associated churn rates.

A market that has been around long enough to provide some useful lessons is the home television market. In most regions, the market for domestic television first evolved through local VHF/UHF transmitters or through what is known in the United States as network television. When cable television systems enter such a market, they see the total market as the sum of the networks' market plus the new business they expect to bring in through a much larger number of television channels, in the process creating market segmentation for the first time through specialized channels. The cable systems' success depends on a minimum number of free over-the-air users who are willing to switch and pay monthly subscriptions for their cable service, primarily for the specialized channels and in many cases for better reception quality. If this process for the entry of the cable systems is reasonably successful, there is a net increase in the size of the overall market pie and the associated advertising revenues.

Taking this scenario one step further, let's now examine the impact of the entrance of a direct satellite broadcast system or DTU-TV into this hybrid market of network and cable television customers. After some marginally successful attempts, significant and measurable impact was made by satellite-based systems only when they could provide high-quality channels in sufficiently largely numbers, not only in underserved rural areas but also right in the markets dominated by cable systems. This has led to the current mature stage of the television market wherein three different media—the networks, cable, and satellite systems—provide competitive services to largely overlapping audiences. How long will this plateau last? On one side, there is uncertainty about the future conversion of VHF transmitters to high-definition television (HDTV) or multiple digital channels; on the other, there is the real prospect of optical fibers eventually reaching most homes, thus unleashing huge bandwidth for a range of multimedia services. The satellite systems too are making big bets on Ka-band, hoping to hold on to their shares in the new mix of the future.

Apart from television, the other growth areas for the satellite medium are radio, broadband, and possibly mobile [6]. In all of these markets, there are competitive media, and the same basic principles for market assessment apply.

Growth Through Integration and Consolidation

We have so far considered the overall market size potentially increasing by the addition of new media and technologies. From the perspective of an individual entrepreneur, there is another way of increasing business. This is by expanding her activities in the overall value chain, as briefly touched on in Chapter 1. Such value chains can provide opportunities for growth through different types of integration and consolidation. Figure 4.2 shows a television value chain, adapted from a more generic value chain described by Wickham [3]. As he explains, in principle, there are two basic approaches for expanding the role in the value chain.

Horizontal integration means enlarging the size of the market share through the acquisition of a competitor. Assuming such an action is allowed by regulators (which generally means that there is still credible competition in the marketplace even after the proposed merger), such integrations are only successful if they create synergy of one kind or another. Synergy can be realized in a number of ways. It could mean pooling of capital-intensive resources (e.g., fewer satellites to procure). Or, it can enable standardization of equipment, thus lowering costs for all, including customers. While it is not always the case, too rapid an increase in size through acquisitions can create cultural problems within the organization, leading to widespread inefficiencies and demoralization.

Vertical integration can be of two types, as shown in Figure 4.2. Backward integration is when at least one supplier in the chain is acquired. Similarly, forward vertical integration refers to the acquisition of follow-on activities. Either type of such integration is beneficial if it makes the interfaces between successive blocks of activities more efficient. This process can, of course, be carried out as far as the ultimate one-stop shopping scenario, providing so to speak the *cradle-to-grave* service. One example could have been the recent merger of America Online (AOL) and Time Warner, wherein the largest Internet connectivity provider, AOL, merged with one of the largest content providers, Time Warner. Unfortunately, subsequent events unfolded quite differently, and the

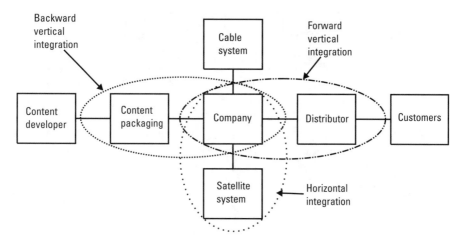

Figure 4.2 Horizontal and vertical integration. (*From:* [3]. © 2000 Pearson Education Ltd. Reprinted with permission.)

merger has not so far produced measurable synergy. In general, the key to success in such consolidations is the talent of the management team. If the top management team does not simultaneously broaden its talent, the chances of failure increase.

Closer to home, a good example is to review how the television services using satellites have evolved, as captured in Figure 4.3, which was adapted from Figure 1.2. For a considerable period, for a combination of regulatory reasons and technological limitations, the role of the satellites was limited to providing a pipeline from the content provider to the local transmitters. While such a *pipeline* role provides steady revenues with modest risks, it captures only a small portion of the overall end-to-end revenues of the value chain.

The first attempt to expand the value chain was to try only a forward vertical integration by leapfrogging the local transmitters and going directly to the eventual customers. This was not commercially successful, largely due to a limited number of channels and no control on the content provision. Once the number of channels was increased, commercial success is being achieved in two broad modes: forward vertical integration only, as in (b), and both forward and backward integration, as in (c).

Competitors

A thorough analysis of the competitive landscape should be based on all of the providers for the end services and not of only those entities within the same medium or technology being targeted. As an example, an enterprise planning to providing television broadcast via satellites has to look carefully at the strengths and weaknesses of all potential competitors providing such services and not just

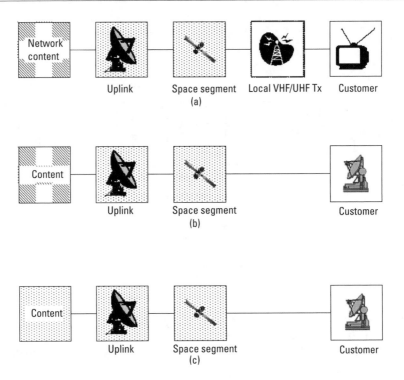

Figure 4.3 Progressive vertical integration in television services: (a) distribution mode, (b) direct broadcast without content control, and (c) direct broadcast with content control.

those using satellites. In an open and balanced marketplace, the consumers are only interested in the end service and generally have the savvy to weigh the relative merits of all providers across the different means of providing such services.

Once all of the competitors have been identified, a thorough and dispassionate analysis must be carried out to identify and confirm whether there is indeed room for another provider. A positive decision can be based on a combination of genuinely unfulfilled demand as well as on the attractiveness of the technology being offered by the new entrant. One common and often costly mistake new entrepreneurs make is to underestimate their own costs in comparison with the established providers. This can very often be due to incomplete appreciation of all costs involved in the early startup phases. Frequently, the founding members fail to recognize that once the project gets off the ground, several of the activities move from the basement/garage model to full-scale operations that involve significantly higher costs. Often there is a mindset that because their new technology is superior, other entities up and down in the value chain would be willing to absorb some of the costs in order to gain entry in their teams. A safe and mature approach is to thoroughly identify all cost centers in a fair and totally unsubsidized manner, if necessary with the help of outside experts with operational experience.

What has been said for costs is even more applicable to potential revenues, as these are often largely beyond the direct control of the entrepreneur. Once again, it is very risky to assume that the competitors would be willing to "play dead" and let a new kid in the block eat their lunch with his fancy new technology. One of the most difficult periods for a company is the first months and years just after the service is introduced. More often than not, the revenue streams take longer to materialize, the costs are rising, the competitors are holding onto their share with fresh discounts, and the investment community is rushing to judgment—or so it seems!

Regulatory Environment

For all satellite-based systems, particularly for those starting a new enterprise, regulatory environment is often the first major hurdle to be crossed. Depending on the type of services planned, such assessments are often required in four categories: spectrum, service, ownership, and international systems.

Spectrum Considerations

Availability of the necessary spectrum over the market area is of course an essential requirement for any satellite-based project. It is not surprising, therefore, that competitors already in the field often resort to creating strong barriers to entry for new entrants by capturing as much of the available spectrum as possible and at the earliest opportunity.

The severity of spectrum-related hurdles a new entrant may encounter depends to a certain extent on the types of services planned and the preferred mode of operation. If the planned service can be provided via leased transponders, in most parts of the world it is feasible to obtain such space segment resources at competitive prices from multiple operators at C- and Ku-bands. To a certain extent, this may also be the case for television broadcast, except for the developed markets where all of the "planned" subbands or transponders for such services are already in use. If a complete satellite capacity is an absolute necessity, the options in most markets are either acquisition or merger with an existing C- or Ku-band operator or utilizing Ka-band if that is appropriate for the planned service.

For mobile services and digital radio, operating at L- and S-bands, the applicable constraints are somewhat different. For digital radio, the overall orbital capacity is relatively quite small in view of the extremely small user antennas with wide beamwidth. Therefore, the barriers to entry can begin to apply fairly quickly once one or two systems begin operations in a region. Mobile systems also operate in the same bands and hence have similar constraints. However, they are resorting to a very high degree of frequency reuse in order to increase the overall capacity.

In summary, in the strategy-development process, the viability of most options is likely to be closely interlinked with spectrum availability over the target markets; such issues should therefore be tackled very early, bringing to bear all of the relevant technical knowledge and regulatory expertise.

Service Policies

Most countries have regulations that govern provision of services of different kinds. In the case of satellites, initial reservations about broadcasters directly reaching the homes, possibly with messages with political or religious implications, delayed the introduction of such services. In some countries, certain types of advertisements, such as for liquor or cigarettes, are not allowed. There can be also requirements for satellite systems to cover rural and other underserved areas.

Ownership

Most countries have restrictions of some kind on foreign ownership when it comes to practically any type of broadcasting. There can also be provisions to encourage part ownership by persons from certain specific segments of the society.

International Systems

All of these factors get multiplied severalfold for international systems, particularly if some kind of broadcast is involved. Fortunately for satellite systems, this difficulty was recognized almost at the outset of the evolution of this technology and viable legal frameworks were established. These led to the creation of INTELSAT and subsequently several other similar organizations, such as Inmarsat and Eutelsat. Member nations of such consortia provided what have come to be known as *landing rights* in their countries from the satellites owned jointly by these consortia. In exchange, all nations, big and small, were assured a responsible management of not only their investments but also the content beamed over their land masses.

The critical importance of these agreements was recognized by other private entities as they established similar links. Notwithstanding the value of such agreements, they have by now all been abandoned as part of the global privatization of telecommunication and related assets.

Internal Environment Factors

These are essentially the strengths and weaknesses of the company and its management team. For an existing company, a new enterprise can draw upon the resources throughout the company provided there is top-management support.

Often there are overt or covert rivalries between different divisions and groups, and it takes sustained effort to bring to bear the best possible resources on the new project.

For a new enterprise, the strengths and weaknesses are those of the management team and its financial and strategic backers. There can be a large variety here. On one extreme, the management team has the experience and wisdom to understand its limitations and acts accordingly. On the other extreme can be overenthusiastic entrepreneurs, who are so enamored by their "technology" that they only pay lip service to other aspects of the business. In fact, the classic saying seems to keep on proving itself in such situations: "they do not know what they do not know" and often their sense of hubris prevents them from searching for areas of their weaknesses in terms of understanding the business. Warren Buffet, one of the most successful investors of our times, drives home the importance of a good management team when he says, "Buying a retailer without good management is like buying the Eiffel Tower without the elevator" [7].

As we have already seen—and will continue to see throughout the book—success requires many disciplines working together in a coordinated manner. This is essentially the genesis of the IBSP. A team with the maturity and sense of responsibility to ensure that all of its components are adequately addressed has a much greater chance of success.

Intellectual Property

In the current highly competitive, and unfortunately overly litigious society, it is becoming increasing critical to protect the company's future through appropriate legal safeguards in the form of patents and copyrights. One of the worst situations a new enterprise can find itself in is that by the time it has obtained the funding and done successful field trials, an established company providing similar service and feeling threatened by the new venture has quietly put up legal barriers through patent applications.

A new enterprise in search of funding has to peddle its ideas to investors of all shades. In this process, it is not inconceivable that the ideas and approaches of the new company can be hijacked by others who have adequate funding to rush to the gate much earlier. It is therefore very prudent that the very first funding be devoted towards protecting the intellectual property.

Strategic Choices and Risk Analysis

With a well-defined mission statement in hand, and a good grasp of the environment in which the business has to operate, the management team is now equipped to identify a short list of viable strategic choices. Each of these choices is likely to have different sets of attributes—some obvious and others somewhat latent. One of the temptations that the management team has to work hard to resist is to push personal favorites. Rather, a careful and dispassionate analysis should be carried out to pick the winner. A good way to start this pruning process is to carefully develop and profile relative risks involved in all options on the table.

Risk Factors in Satellite Systems

Ventures involving space technology have traditionally been considered riskier investments compared to other technologies in telecommunication and broadcasting. This is primarily due to the technical risks of launching a satellite or similar payloads into space—the satellites not only account for a major part of the capital investment, but under present-day technology, once the spacecraft are in outer space, they cannot be repaired or even serviced.

Because of this overriding and *visible* risk in satellite systems, it is not uncommon for all other risks to receive less attention than they also deserve. The fact remains that, like most other businesses, there are risks beyond technical risks for satellite-based systems. In fact, we can recall that several of the expensive lessons captured in Chapter 2 arose from nontechnical risks. What is needed is a balanced risk management discipline that permeates all activities in a coordinated fashion. The following paragraphs provide an overview of three major categories of risks: technical, market, and organizational/financial.

Technical Risks

Satellite and launch-vehicle engineers can justifiably claim to have developed sound principles and methodologies for identifying and retiring risks at all stages of such programs. Their painstaking efforts and procedures have progressively improved the reliability and availability of in-orbit assets. Once the satellites have successfully been launched in their appropriate orbit, most of them operate satisfactorily for extensive periods that are no longer controlled by hardware failures, but rather by the amount of consumable stationkeeping fuel onboard. Concurrently, the success rate of launch vehicles has also continued to improve.

This success in controlling and managing technical risks is even more credible when it is recognized that concurrently newer technologies have been introduced to enhance the in-orbit capabilities in a variety of ways. This has been essentially achieved through a judicious mix of rigorous qualification requirements at all stages and redundancy and other fail-safe features throughout their products. A single-point failure is an anathema for engineers of any stripe in this field.

An inherent component of technical risk is schedule risk, particularly for the space segment. Any new technical risk identified during the program often requires additional tests, delta qualifications, and sometimes new components, materials, or designs. Of course, schedule risk can also translate to market and financial risks in a majority of cases.

However, several satellites have in recent years encountered in-orbit failures to a much larger extent than before. It is debatable whether this is a result of excessive cost cutting, pushing the envelope too far too fast in terms of spacecraft capabilities, or a combination of both factors. The net result has been an extraordinary increase in insurance rates. These issues are serious and are leading to major efforts for quality improvements and increased involvement by several customers in the monitoring of spacecraft programs.

Market Risks

As we have already seen in Chapter 2, market risks can be more critical for DTU systems than for the more traditional satellite systems. The salient points are recalled here:

- Transponder leasing still accounts for a sizeable part of the total global space segment. Thanks to a fair degree of standardization, such transponders can switch from one type of service to another. Therefore, once a particular orbital location has acquired a critical mass in terms of antennas accessed, the satellite operator tends to build space assets to fill the available spectrum at that location as efficiently as possible rather than to fit a certain demand model.

- For DTU consumer business, market data can be much more critical. Of course, its criticality plays out in different ways for different DTU services. In most of the DTU services, satellite systems are competing with other technologies. In such situations, the market share that can be captured is not only price sensitive but also time sensitive.

- For satellite broadcasting systems, the satellite size and cost do not increase with an increase in the number of viewers or listeners. Instead, it is entirely controlled by the number of satellite channels broadcast, receiver sensitivity, and coverage area size. In that context, market data can have a time sensitivity when seen in the context of what competition is providing. As new entrants come in, their content and extent of local channel broadcasting have to be matched. As cable systems switch to digital operation, for example, they begin to match qualities and start offering bundled services that cannot be easily matched.

- For the new and still evolving satellite radio broadcasting market, the successful emergence of competition and alternative technologies (e.g., IBiquity) can create new elements of market risk unless the system has enough agility to match the new offerings.

- For two-way broadband satellite systems, the capacity needs for the millions of return uplinks coming from small dishes, if not properly anticipated, can well become risk factors for the market share.

Financial Risks

To a degree, most technical and market risks translate into financial risks. Technical risk surprises can lead to increased material and manpower costs, ad hoc architectural changes, and schedule slippages, with some or all of which leading to increased financial risks. Even in fixed-price contracts, while the increased costs are to be borne by the contractor, the schedule slippages can translate to market risk for the customer.

Sometimes, the technical risks impacting the program can be completely out of the control of the customer and the spacecraft contractor. A common example is launch failure in a different program, which can often lead to schedule delays and increased insurance rates for all parties using the same launch vehicle series.

Market risks flow through to financial risks much more directly. If the target audience size is not realized according to the business plan time profile or the target market share is captured by another competitor, the resulting decrease in revenues can lead to serious financial problems.

Of course, there are risks within the financial management of a project as well. A common financial risk is overspending. This can arise from a variety of causes, such as unexpected inflation, poor or inadequate competition in procurement processes, and top-heavy administrative and monitoring functions. Financial risk can also arise from poorly structured financing terms, which are more susceptible to changes in external factors in the financial world.

Risk Profiling

Figure 4.4, adapted from [8], is a good example of bringing together different types of risk factors in order to develop the overall risk profile for a particular option. In this figure, organization risks are also included as part of the box on financial risks. The three boxes list some representative factors that may apply to typical options. However, different options can and often do have different sets of factors in different categories of risks. Such profiles, with as much quantitative backup data as possible, can be useful in comparing different strategic options and choices.

Narrowing Down the Strategic Choices

Once the external and internal environment analyses have been completed and the risk factors thoroughly analyzed, the stage is set for the formulation of a

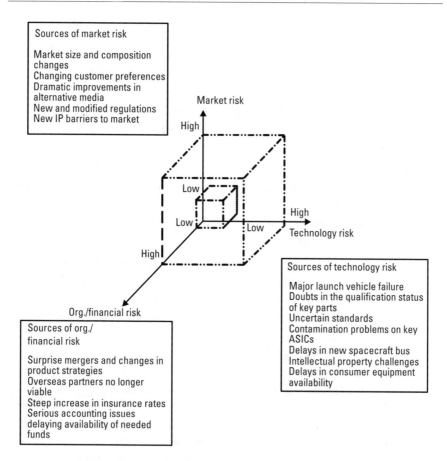

Sources of market risk

Market size and composition changes
Changing customer preferences
Dramatic improvements in alternative media
New and modified regulations
New IP barriers to market

Market risk

High

Low

Low

Low

High

High

Technology risk

Org./financial risk

Sources of org./financial risk

Surprise mergers and changes in product strategies
Overseas partners no longer viable
Steep increase in insurance rates
Serious accounting issues delaying availability of needed funds

Sources of technology risk

Major launch vehicle failure
Doubts in the qualification status of key parts
Uncertain standards
Contamination problems on key ASICs
Delays in new spacecraft bus
Intellectual property challenges
Delays in consumer equipment availability

Figure 4.4 Risk profiling. (*From:* [8]. © 2000 John Wiley & Sons, Inc. Reprinted with permission.)

short list of strategic choices. These choices should be developed through a team process where no preconceived direction or favorite choice is imposed from the top. Rather, the process should let the chips fall where they may. In an extreme example, if the answer is that a satellite-based solution is not the right answer, that should be accepted and a mature decision made whether to exit the market altogether or to adopt an alternative solution.

After a short list of viable options is available, the top choices should be critically examined through a set of filters of the type suggested by Wickham [3]. The following paragraphs illustrate the use of these filters from the perspective of satellite-based systems wherever applicable.

- *Consistency.* The option should be consistent with the objectives of the mission statement. It should be challenged to deliver the planned set of services to the chosen market segment *at the target prices.* The temptation to implicitly modify the objectives due to a fascination with a technology or market should in general be avoided. If what is learned in this process is indeed attractive as a business strategy, the mission should be revisited and revised.

- *Attractiveness.* Will it provide the minimum return on investment planned and within the total investment planned? Will the services offered attract the audience in the planned numbers?

- *Acceptability.* This filter goes beyond attractiveness. Will the current and future investors and stakeholders accept it as a viable business with which to be associated? Will it attract the right talent as employees? Will the suppliers in the value chain accept it as a basis for becoming a long-term partner?

- *Feasibility.* This includes not only technical feasibility, but also the limitations of the management team in terms of the depth and breadth of its capabilities. A positive score against this filter should be predicated on feasibility of building the system *within the market window time frame.*

- *Validity and vulnerability.* This captures the importance of underlying assumptions applicable to all aspects of the program. In essence, how sensitive are the chances of succeeding to the underlying assumptions? What happens if the market forecast turns out to be wrong? How fungible is the system to switch markets halfway through implementation and recover at least part of the investment? In the context of satellite systems, how dependent are the major components on the completion of another project (e.g., qualification of a new spacecraft bus, success of a new launch vehicle, or acceptable yields of advanced ASICs within the power consumption for the receivers).

Selected Business Strategy

The final outcome of the part of the IBSP discussed in this chapter should provide a plan of action to meet the objectives of the mission of the enterprise. That is what a sound business strategy is supposed to do. Obviously, there is no sure way to absolutely predict the future; however, a good management team minimizes the risks of failure by following a systematic process of the kind described in this chapter and by learning from past experiences, both good and bad.

The selected business strategy will form the basis of a business plan, which we will address in the next chapter. Before we do that, we list a few specific questions relevant to our industry that should have been conclusively answered by now:

1. How reliable is the market forecast? Is it likely to change downwards with time or with the emergence of alternative means of providing the same or similar service?

2. Is the market forecast quite sensitive to the price and subscriptions to be borne by the ultimate consumer? What is its sensitivity to higher prices in one or both of these categories?

3. Will your system meet the price targets? What is level of uncertainty in your cost estimates for the total infrastructure and operations?

4. Does the consumer equipment exist in the market? If not, will it be proprietary to your system? What is your confidence level that you can meet the price and schedule targets for this equipment?

5. What percentage of your satellite is based on available production lines, and how much has to be custom designed for you? What is your confidence level in the cost estimates and schedule reliability?

6. What percentage of your overall system is fungible and can be utilized for other services if your markets do not materialize to the extent planned? Do you have fallback options?

7. Are you planning to vertically integrate the entire operation from content to customer service? Will this be based on partnerships with established firms or are you planning to create the entire chain in house?

8. Are there any regulatory uncertainties still unresolved? Can they become show stoppers midway through the program?

9. What is the competence of your founder-management team? Is it overly satellite centric? Is there sufficient importance given to the ultimate consumer in your internal decision-making process?

Before closing, we present two case histories for discussions around the principles presented in this chapter. The first case history documents how sometimes a high risk pays off. However, its outcome was not known in the beginning. The second case history summarizes the background for a project that has been discussed for a considerable period. The summary poses some questions for discussion.

Case Studies

Case History 4.1: Rene Anselmo's Gamble

It was the early 1980s, and the world of international satellites was dominated in most parts of the world by Intelsat, an international consortium of nearly 100 countries at that time. Several nations and individual companies were domestic systems; however, it was very hard to penetrate the international markets because the INTELSAT's owner nations were also generally its customers as well.

Rene Anselmo, a successful broadcaster, was determined to break what he considered to be a "market stranglehold" of treaty organizations and their signatories. He managed to get a license to operate a separate satellite system and risked his personal fortune to build it. Within the limited funds he had, he decided to start with a bargain basement single satellite and signed up to launch it at an equally low rate offered by a brand-new launch vehicle, Ariane. Fortunately for him, the launch was successful and thus was born Panamsat, currently the world's fourth largest international satellite system.

As a business strategy, Anselmo's was a high risk one from several considerations. The marketplace was largely controlled by government-owned telecommunication operators who had vested interests in INTELSAT through their long-term investments. It was equally risky to start with a single satellite and then launch it on an unproven launcher in exchange for a low promotional price. However, the payoff was that of a complete value chain from customer to customer rather than revenues from only part of the chain, such as leasing transponders.

Anselmo's strategy worked partly because he was lucky with his first launch. His dogged determination (pun intended with his Dog Spot letters and commercials!) did accelerate the opening up of the marketplace to private entrepreneurs all over the world.

Looking back, was the Anselmo business strategy correct or too much of a roll of the dice? Did he properly take into account all environmental issues? Were there other alternatives?

Case History 4.2: Iridium System

It was the late 1980s, and the first optical fiber cable TAT-8 had just started operating across the Atlantic, challenging the so-called complacency of the satellite systems market leaders, including INTELSAT and Inmarsat. The manufacturing side was also dominated by a few long-term incumbents led by Hughes Aircraft and a few others.

Motorola was a giant in the mobile phones field but had no visible presence in the satellite field. It was time for a bold entry, and Motorola did just that with a revolutionary proposal to build a global mobile satellite system using a large number of satellites in LEOs. The unique attribute offered was much smaller time delays and instant connectivity via mobile phones literally from anywhere to anywhere on the surface of the Earth.

The process of building the system was as historic and trail blazing as the system concept itself. Regulatory approval was obtained from the International

Telecommunication Union (ITU) after a major campaign at government levels and funds were raised internationally, with the dual benefit of gaining access and spreading the risks. The construction of the system with 66 satellites interconnected by direct links was begun in 1990.

After more than seven long years, the system was finally completed, establishing several novel advances in system design and satellite manufacturing in the process. The user equipment took a little longer to arrive, and, when it did arrive, it was unfortunately too bulky, very costly, and initially had serious performance issues.

The story from then on was all downhill—on a steep slope. The more than $5-billion, truly revolutionary system filed for bankruptcy in 2000. It is now operating on a much smaller scale, mainly for strategic government services. There are no public plans to build the replacement when the satellites begin to die around 2013.

So what went wrong? Most of the satellite community came with a quick diagnosis: LEO systems were no good and needed too much money to get started. This was chilling news indeed for the satellite and the launch industries, as they had expanded their capabilities severalfold in anticipation of additional new multisatellite nongeostationary systems. This excess capacity is even today haunting the industry, exacerbated by the more recent telecom meltdown.

The spectacular rise and fall of the Iridium system will obviously be debated for a long time to come. It is unlikely that there will be consensus on precisely why it failed. This is mainly because, for a complex system, failure just like success is very hard to pinpoint to just one or two factors. The teams that conceived and implemented the project were a group of dedicated and experienced individuals and companies around the world. In author's opinion, despite Iridium's failure, there are a lot of merits in LEO-system concepts. Discarding LEO systems would be equivalent to the proverbial "throwing the baby out with the bath water." The following questions are provided to stimulate a discussion in the context of this chapter:

- Was the system development process too satellite centric? Did it not give adequate importance to the consumer equipment first? Had it done so, would the outcome have been substantially different?

- When the Iridium project started, the cellular industry was still in its early stages of development. If the Iridium system had gone into service much earlier than it did, with competitive user handsets, would that have altered the final project outcome?

- Would a much lower cost of service have created a critical market niche for this system?

- Should future satellite mobile systems abandon the LEO approach due to its high starting investment?

References

[1] Gray, C., and E. Larson, *Project Management: The Managerial Process,* 2nd ed., New York: McGraw-Hill, 2003.

[2] Aaker, D. A., *Developing Business Strategies,* 2nd ed., New York: John Wiley & Sons, 1988.

[3] Wickham, P., *Financial Times Corporate Strategy Casebook,* London, England: Pearson Education Ltd., 2000.

[4] Wickham, P., "Developing a Mission for an Entrepreneurial Venture," *Management Decision,* Vol. 35, No. 5, 1997, pp. 373–381.

[5] Weaver, D., "MobileMemory," presented as class project on project management, George Mason University, December 2003.

[6] Sachdev, D. K., "Three Growth Engines for Satellite Systems," *AIAA 20th International Communications Satellite Systems Conference,* Montreal, Canada, May 12–15, 2002.

[7] Buffet, W., Chairman's Letter (1995) at http://www.berkshirehathaway.com/letters/1995.html.

[8] Day, G. S., and P. Shoemaker, *Wharton on Managing Emerging Technologies,* New York: John Wiley & Sons, 2000.

5

Business Plans

> If you want to make God really laugh, show Him your business plan.
> —Barry J. Gibbons

Surprising as it may seem, this quote is the title of a recent e-book on business planning! The book catalogs a number of spoofs and mistakes that entrepreneurs often make while capturing their ideas and concepts in the form of a business plan document. While a lot can be said about the average quality of business plan documents, or lack thereof, we have to recognize that the acceptance or otherwise of a typical business plan is dependent not just on its veracity, depth, or underlying strategy, but also on its *timing* in terms of the prevailing general business climate, or macro-environment, as we discussed in the previous chapter. As is well known, in the heyday of the dotcom explosion, a "business plan" drawn on a napkin was often deemed adequate to deliver millions of dollars of investments overnight. However, once the venture-capital marked soured, even the most comprehensive plans had to struggle just to get an audience for an initial briefing.

Perhaps these widely different macro-environments were two extremes in the highly unusual business cycle at the turn of the century. However, even in normal times, it is often hard, even in the most entrepreneurship-friendly parts of the world, to get started on a business idea. Why is that? While obviously there is no one unique answer, the most common reason is that, given the ever-broadening range of technologies and services, the management teams behind an otherwise viable concept do not always capture all pertinent aspects in a logical manner, either in their business plan document or in their face-to-face presentations or both. This is true not just for new enterprises. Even within established corporations, especially those that have grown over the years largely through internally generated resources, similar difficulties are quite common. Often, they put together the plan document in a hurry in the management floors and then parade their otherwise highly capable technologists in front of investment community ostensibly to impress them. Unfortunately, such experts are often not even familiar with the contents of the document; even if they are, they are not always able to present in a clear business language their complicated concepts to external audiences and often even to their own top management!

Purpose and Objectives of a Business Plan

While certain common and essential elements must be included in practically any business plan, its flavor and depth can vary quite a bit depending on what it is supposed to achieve. Its contents are also heavily influenced by the stage at which the company is (e.g., whether it is it a startup, an established enterprise, or a new startup division within an established enterprise).

The first step toward an effective business plan is to clearly identify its purpose and objectives. While these can cover a wide range, the most common objectives of a business plan are:

1. Raising capital;
2. Attracting partners;
3. Motivating and synchronizing company staff.

The need for investment capital, of course, arises at practically all stages of a company's evolution. At the startup phase, while the capital needs are relatively small, they are generally harder to finance. This is because the investors and markets do not know the company, and often there is no directly relevant past record to give them adequate confidence to agree to get locked into what could end up being a very risky enterprise. This proverbial "chicken-and-egg" syndrome requires not only a different kind of a business plan document, but also a much higher degree of personal rapport and confidence-building efforts by the founder/management team. For an established company, raising additional capital is relatively easier, provided of course the past business track record until that point is acceptable and positive. If the project needs so permit, an established company may also have the flexibility to defer additional funding efforts until the appropriate macro-environment conditions are in place. However, such efforts can face bigger hurdles than even by an unknown startup if the past performance, in terms of timeliness, proper accountability, and returns on investments, are not up to the mark.

Satellite-based systems tend to be upfront capital intensive, and in most cases most of the capital gets sunk long before the very first revenue streams begin. This is often a serious hurdle for a startup, particularly for those venturing into new and unproven markets. Chapter 2 touched upon some of the consequences of these factors under adverse circumstances. In general, a system architecture that manages to stagger the investments of a startup company through incremental buildup of the needed assets will find it easier to attract investors.

The second or alternative objective of a business plan can be to attract additional partners. Such partners can be from a wide range of backgrounds and experience. If the founding team members are largely finance, market, or operations oriented, the enterprise could be looking for technical managers with proven records of implementing complex programs. While locating such experts is generally not difficult, persuading them to accept the challenge and uncertainties of a new and often partially funded project, in lieu of possibly the relatively secure environment of a large corporation, is another matter altogether. In such situations, the presentation accompanying the business plan document assumes higher importance. Generally, experienced engineering managers look for a mix of intellectually challenging opportunities and plausible potential for a lucrative and secure career within a financially sound project. Their scrutiny of the business plan document itself may be largely focused on the attractiveness of the products and services and on their underlying technology.

The opposite scenario is more common, wherein a couple of mid-level and technically smart managers are the startup founders who have decided to take the risk of leaving secure careers in an established company in order to finally build their dream products or services. They are generally looking for partners who will manage the finances and take care of day-to-day operations. Their business plan, in order to successfully move toward the goal of attracting such partners, needs to clearly demonstrate the practical and financial viability of the dream products or services. Specifically, they run the risk of not securing the right partners if the business plan is heavy on the novelty of the founders' technology but does not adequately address the inherent risks of feasibility, delay, or cost overruns for the project.

For satellite-based systems, founding entrepreneurs of either category often look for one or both kinds of the integration discussed in the last chapter. An example of partnership with a downstream entity would be with automobile manufacturers for the original equipment manufacturer (OEM) addition of satellite radios. For such partnerships to be successful, the concerned companies have to be prepared for extremely rigorous trials and some kind of risk sharing.

Another example of partnership in the form of backward vertical integration would be with the suppliers of the satellites and other infrastructure. Such partnerships, more common during lean industrial workloads and with startups, can have mixed results. On one side, the satellite or ground segment manufacturer is likely to have the incentive to maintain a greater cost control and to deliver within schedule, as his or her own investment is also at risk in the enterprise. On the other side, a customer beholden to his or her own suppliers for a good part of the finances can lose some of the oversight authority crucial to maintaining performance standards at all stages of the program. Once again, some of the problems discussed in Chapter 2 can be traced in part to such factors.

The last objective of a good business plan is to motivate employees. To bring them in sync with the company's vision and mission is a very desirable objective, although it is rarely the driving objective and is often overlooked altogether. As discussed in the last chapter, the mission statements are often not even known to all of the staff. The same can often be true for the business plan itself. Often the mid-level managers and their staff are pulled into a business plan development cycle at the last minute in order to provide data to convince outside financiers about some claims of performance already being asserted in the plan prepared by top management alone.

Active participation of staff in the preparation and updating of the company's business plan can provide several tangible and intangible benefits in terms of staff motivation and the overall productivity of the organization. Whether a business plan is prepared initially by the management team or in a "bottom-up" exercise by the staff, it can have a positive effect in motivating all of the staff and encouraging them to feel firsthand responsibility for the company's overall success. There is also another tangible benefit. In most technology-based enterprises, the pace of change is so rapid that a business plan can pretty soon become obsolete at its core strategies and objectives. An organization that uses a business plan as a living document and as a motivating and binding agent for its staff also reaps the benefits of having the document current and in step with technological advances within and outside the company.

Business Plan as Part of IBSP

In Chapter 3, we introduced the concept of an IBSP in Figure 3.1, wherein all of the activities for a satellite-based project are carried out in a fully coordinated manner. In Chapter 4, we addressed the major activities leading up to a business strategy in Figure 4.1. Once the company's business strategy has been developed, the next step is the preparation of the business plan in order to meet one or more of the objectives discussed earlier.

In some respects, a business plan is the first point where all of the IBSP activities can and should come together, even though some of the follow-up activities may not have yet started in full force. The extent to which the preparation of the business plan is a truly organizationwide activity can often decide the quality of the enterprise and may in many cases decide the fate of the project and the enterprise itself.

Figure 5.1 expands the part of the IBSP surrounding the business plan. Business strategy is the principal input with several subsidiary inputs from the activities already completed, such as market share and regulatory assessment. Of the follow-up activities, the most critical inputs are infrastructure costs and estimated revenue streams. For a new enterprise awaiting its first significant funding, the internal talent and resources at this stage are generally limited to the management team and perhaps a few consultants in specific areas. Their ability to accurately and convincingly develop and estimate future cost and revenue streams is one of the critical tests of their experience, as we will discuss further under contents of the plan itself.

Success or lack thereof of a business plan should be judged by the extent to which it achieves the objectives set for it—funding the enterprise, finding partners, and motivating the staff. As shown in Figure 5.1, the IBSP can come to a virtual halt if the funding for the building of the system through the follow-up activities is not secured in a timely manner. An equally important output is to provide the system-planning phase with a complete picture of the whole project in one coordinated document.

Business Plan Contents

A responsive business plan document provides a comprehensive picture of what an existing organization's future targets are or what a new enterprise aspires to be. It is therefore extremely important to make this document as coherent and logically structured as possible. If the company or the project is following an

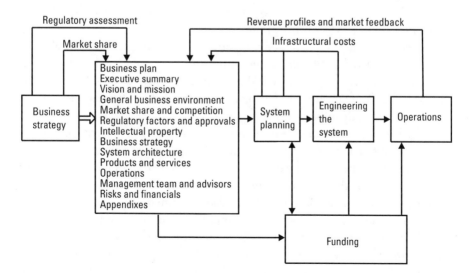

Figure 5.1 Business plan formulation.

integrated process, such as the ISBP developed in this book, preparation of the business plan can to a very large extent follow the structure and sequence of this process, with some additions and differences of emphasis in individual areas, depending on the plan objectives. It is interesting to note that a group effort for preparing the business plan document itself can often lead the company to adopt a structured process such as IBSP or to make improvements to its existing processes. We will now address each of the major sections of a business plan with emphasis as relevant on satellite-based enterprises.

Executive Summary

This is generally the most critical part of the plan. Like all credible summaries, it must capture the substantive points in the full document in a brief, equitable, and succinct manner. Furthermore, it should be kept in mind that during initial dialogues with perspective investors and partners, this summary may be all that they would look at. Therefore, the executive summary should be like a stand-alone miniature business plan.

The executive summary should clearly convey the key objectives of the business plan discussed earlier, be it financing, finding partners, or simply making organizational improvements. Reading the summary itself, an experienced reader should be left with little ambiguity about the company's business strategy and its objectives. In other words, it is the summary that upfront acquaints the reader with what the company wants to be in the near and long-term future.

While it is important to underscore the targeted market share, the credibility of such numbers in the reader's eyes depends not only on the methodology adopted but also on a demonstrated awareness of direct or indirect competition to the product or services being proposed.

Relevance and ownership of any critical intellectual property can often be a plus, particularly if differentiation with current providers is being claimed. Along the same vein, any major weaknesses or risks involved should be explicitly stated in the summary itself rather than being relegated to a small print in the bowels of the full document.

Unless the audience already knows the company, one of the significant positive results of a good executive summary is a growing confidence in the management team. Experienced investors often attach more importance to the caliber and experience of the top management than to the proposed technology. The summary document should highlight the most pertinent experience and qualifications of the management team and, if important, for their key advisors as well.

In view of the large capital investments necessary early in most satellite-based projects, it is very important that credible estimates are provided in the summary for the total investments needed and not just for the initial sums needed to get started. Equally important is the provision of an easily understandable summary of the estimated revenue streams and the return on capital and other financial benchmarks likely to be of interest to the audience. The revenues figures in all cases should flow from the stated market forecasts and should include honest estimates on their achievability.

Finally, a good executive summary, accompanied by a clear and credible presentation, can go a long way in opening new doors for the business. Above all, for investments to flow back through such open doors, the executive summary should be brief, focused, persuasive, honest, and factual [1].

Vision

In earlier chapters, we started the IBSP with the mission of the company. However, in the business plan document, which will be distributed to the investment community in general, it is often desirable to start with a brief enunciation of the vision behind the enterprise.

What exactly is vision? A novel but effective way to illustrate what a good vision should be has been given by Wesley Truitt [1] by recalling a memorable event from the political and social world. This is the speech by Dr. Martin Luther King, Jr., on August 28, 1963, at Washington, D.C., "I have a dream that one day the nation will rise up and live out the true meaning of its creed...all men are created equal." While this was obviously a political oration and not a business venture announcement, as a vision statement it was clear and focused on its objectives and inspired others to achieve the objectives. Vision is not the more detailed mission statement that follows or the strategic plan that is based on these statements. As Wesley Truitt so aptly points out, Dr. King would have lost all of his impact if he had said instead, "I have a strategic plan..."

Vision is a broad, but clear, statement by the management on the overall strategic direction the company wishes to pursue. If it is a startup, the vision can be an expression of its intrinsic aspirations; if it is an established company, the vision captures for the outside world as well as for its staff where it wants to be in the long term or by a specific target date or year. Often there is a tendency to pack multiple objectives in a vision statement. Apart from appearing as a document created by a committee, such a statement can often confuse the audience. For modern satellite-based systems, the brief vision statement leading the business plan should enunciate in clear language the role of the company in the sector and its relationship with other sectors.

Mission

Following the vision statement, the business plan explains what the mission of the company is planned to be. Unlike the vision statement, the mission statement has to be specific and should capture the targeted role for the company, who the customers are, and who the stakeholders are in the company, as explained in Chapter 4.

In the satellite sector, the mission statement, for example, should clearly identify whether the company wishes to be a service provider to the end customers or a bandwidth and coverage provider to other local and regional entities. Established operators, particularly with business on a global basis, often are in both camps. However, a new enterprise is generally better off in the financial world if it starts with a clear and simple mission.

Environment Analysis

Chapter 4 addressed the basic principles of environmental analysis as means of converting the mission statement to a viable business strategy. The business plan should capture all of the key elements of this analysis with more specific reference to the planned business model. Common elements generally applicable to satellite-based systems are addressed now from these perspectives.

General Business Environment

The business plan document should go a little beyond the principal macro-environmental factors identified in Chapter 4, such as prevailing interest and inflation rates and taxation policies. It should also address the overall sector environment applicable to the proposed business.

As we saw in Chapter 1, the total satellite sector had a global turnover of nearly $100 billion annually during the 2002–2003 time frame. This is the sum of all products and services up and down the value chains. However, as the more detailed component charts show, there were positive and negative trends that could be of relevance to the proposed business. Thus, while the revenues for the direct-broadcasting services were demonstrating a healthy growth, the transponder business appeared to be a bit stagnant. Similarly, the manufacturing sector, particularly for spacecraft and launch vehicles, demonstrated certain trends. The proposed business model in the plan should be consistent with such sector-wide trends. If there is a strong rationale to deviate from such data, this should be explained in the plan.

The plan has to take into account any recent events, good or bad, in the satellite or related sectors that could have a direct or indirect bearing on the overall assessment and investor confidence in the proposed project. Specifically, as discussed in Chapter 2, the satellite sector encountered some of the most far-reaching business failures in the 1990s. While the diverse audience of the business plan may or may not fully agree with the lessons derived in this book (see

Table 2.1), they will certainly apply their own yardsticks or checklists based on these business failures in the sector. Even the much larger failures in telecommunication ventures could impact the proposed project (e.g., in the confidence in the marketing data based on future telecommunication demands). If the proposed business is consumer oriented, direct broadcast, or mobile, it has to show as a minimum an awareness of the past failures and to demonstrate in a professional manner why the proposed business model does not carry the same risks that led to failures in the past. Several lessons developed in Chapter 2 become relevant here.

Market Share and Competition

This is one of the most critical elements of the business plan. With credible and well-researched data, the audience has to be convinced that, given the necessary resources, a share of the identified market can indeed be captured by the proposed project.

If a significant part of the identified market share is planned to be secured through differentiation with one or more incumbents already in the same business sector, verifiable data should be included to justify such a claim. A current example is the provision of broadband and Internet access via satellites. The business plan should clearly identify the criteria on the basis of which a current digital subscriber line (DSL) or cable modem user would switch to the proposed satellite system. Will it be price, access, quality, speed, or upfront and subscription costs?

This example also illustrates the potential weaknesses in market share identification. The extent of penetration of DSL and cable modems for broadband services varies in different parts of the world. In all markets, a certain portion of the potential users is currently not served by either of these media. Before gross numbers for such underserved customers are used as a justification for the proposed business, care must be exercised to further break down such numbers to those who cannot be served due to lack of these media and those who are not yet willing to get high-speed connections at the prevailing upfront and monthly costs. As Chapter 2 pointed out, underserved markets can often turn out to be untenable as well.

For businesses targeting the end users, perhaps the most common omission or weakness in their plan documents is the total absence or more than acceptable ambiguity about the total costs to the consumer. Such a weakness (real or perceived) can dramatically reduce confidence in the plan, given the well-known correlation between market size and user costs for practically the entire consumer industry. Chapter 2 highlighted this as one of the major reasons for the failure of earlier such ventures. It is not adequate for the business plan document to avoid having a flavor of being satellite centric. If the consumer equipment does not already exist in the market place, verifiable and quantitative evidence on the expected cost and timeframe of availability should be explicitly provided in the plan. The market data forming the basis of all financial projections in the plan must be based on these numbers without any optimistic extrapolations or excessive hand waving.

Regulatory Factors and Approvals

Three major categories of regulatory factors were identified in Chapter 4: spectrum, service polices, and ownership rules. All of these, if applicable, should be addressed explicitly in the plan.

Spectrum translates generally to the availability of orbital location for each of the satellites planned to be used. If the project does not involve immediately procuring full satellites, but is instead based on leasing capacity or bandwidth on other satellites, the confidence level in the venture's ability to realize the projected revenues is enhanced if the orbital location happens to be a popular or "hot" with the market. Along the same lines, the capture rate of a consumer service venture (e.g., direct broadcast of television programs) can be expected to be higher if a large number of consumer dishes are already pointed at the same location.

The ITU is the final authority on the allocations of orbital locations at each frequency band to its member nations, generally on a first-come-first-served basis. Such allotments stipulate that the orbital location must be physically used within a specified period, currently nine years. Each such location is in turn authorized for use by the national regulators to companies or service operators. (For international systems, one of the member nations fulfills this role.) The national regulators, such as the Federal Communications Commission (FCC) in the United States, can also stipulate major project or infrastructure milestones as a condition for the authorization of the location. Finally, in some cases, the orbital location or part of the spectrum at an orbital location can be put out for competitive bids through controlled auctions by national authorities.

The business plan should clearly address these aspects as applicable to the venture. Major funding is unlikely to be available without a clear authorization of spectrum or orbital location, or identification of in-orbit capacity proposed to be leased from other vendors. If the enterprise has already secured the necessary authorization, the plan document should substantiate the ability to meet the milestones specified by the regulators, along with corresponding expenditures and deliverables.

In view of the criticality of the spectrum to a satellite-based venture, securing of an orbital location or resource often becomes the very first activity of the founders of the enterprise, often long before the entire business is fully thought through and the associated business plan prepared. Once the spectrum is authorized, utilization of the *entire* spectrum authorization often becomes a perceived necessity, regardless of whether the entire capacity is really needed by the venture. This unintended consequence of the current spectrum authorization processes often contributes to an investment level higher than it needs to be or can be fully recovered through realizable revenues. If there is indeed a likelihood of excess capacity in the system, the plan document should present ways of mitigating its impact through staggered investments or generic fungible designs permitting leasing of access capacity to other operators.

Service regulations can be both national and international. Except for the *unlicensed* bands, in most cases the regulators can specify to varying degrees the services that can be provided in a specified segment of the spectrum. This is particularly rigid for broadcasting authorizations for television as well as radio. For other locations more suitable for transponder leasing, there is more flexibility. For international systems, such regulations can in some cases vary in detail in different countries. The plan document should demonstrate awareness of and compliance with all such regulations.

Ownership regulations are generally resolved at the time the national regulators make the authorizations in the beginning. However, while seeking partners, the business plan should highlight, at least in the verbal presentations, any likely restrictions on the types of partnering allowed by the national regulations.

Another type of restriction may not strictly fall in the definition of regulations, but nevertheless can decide the chances of success for the venture. To a widely varying degree, individual countries have restrictions on the type of content that can be distributed or broadcast in their countries. Some of these restrictions can be absolute (e.g., types of commercials or programming), while others often do get resolved through minimum local ownership requirements. Before the business plan is presented to clients in different countries, it is obviously beneficial to check out any such restrictions.

Intellectual Property

An active patent, approved or awaiting approval, can be a significant strength of the plan, provided the project architecture is a direct beneficiary of such a patent, either in terms of creating a new market segment altogether, enhancing market share through differentiation, reducing consumer costs, or a combination of all of these. Such patents can either belong to the founders or can be licensed from the original inventors.

If the patents are only in the resumes of the founders and management/staff, these can still be valuable, but only in the context of the management team section of the plan. Often, the actual rights of such patents could belong to the organizations for which the staff works. If this fact is relevant to the project, it should be highlighted in this section.

There is another side of intellectual property that can have an adverse impact on the project, often after considerable investments have already been made. This situation can arise where the engineering or system design may turn out to be infringing the patents of other individuals or organizations. A claim by a competitor about possible patent infringement can be a significant risk to the venture, and the management team has to be in a position to demonstrate due diligence in this regard.

Business Strategy

This section of the plan generally brings together all other elements in the plan, in accordance with the principles discussed in the last chapter. The strategy has to clearly demonstrate how the company plans to realize its vision and implement its mission. It has to take into cognizance all of the environmental factors relevant to the project. In other words, all of the items in other sections should not be mere window dressings but rather integral parts of the plan. While the business plan does not necessarily have to list all of the strategic options considered, it could be useful to justify the final choice via comparison with one or two obvious choices that had been considered in equal depth but were dropped for sound business reasons. In this regard, the set of filters and questions described toward the end of the last chapter can be useful.

Finally, the strategy described in the plan document should be one for a company in the real business world and should substantiate how the critical elements in other parts of the plan, such as market share projections, investment profiles, and the anticipated revenue, will make the chosen strategic path a sound business preposition with acceptable risks.

In the satellite business sector, at first glance there might appear to be a relatively few strategic choices from which to pick. Let us briefly consider a few examples to confirm if that indeed is the case.

First, we look at the classic business segment of providing leased transponders to service providers. Here, it may indeed be the case that there are only a few strategic options. From the outset, the most important factor is whether an appropriate orbital location is available. This business sector has traditionally grown around relatively few international and national administrations or companies, who have over the years obtained parking rights for most of the viable orbital locations at the popular C- and Ku-bands. Therefore, opportunities for a brand new entrepreneur may be relatively few unless prospects due to privatization and spinning off of subsidiaries arise. Recently, due to factors beyond this industry sector, some opportunities for buying part of erstwhile international systems have become available for investment. This can be an opportunity for new entrants in the business, especially those with deep pockets. Ka- and other bands do have a large number of orbital locations not yet occupied; however, it is too early to say if these would provide sound opportunities for the classic transponder-leasing business.

Let us now consider a relatively new market segment for the satellite sector—provision of broadband two-way Internet services. Provision of two-way broadband services does indeed provide several choices to an entrepreneur, each with its own tradeoffs of some complexity. To list a few:

1. Ku-band service as a derivative of satellite television broadcasting;
2. Hybrid Ku-/Ka-band service;
3. Ka-band-only service through dedicated satellites, either transparent or with onboard processing;
4. Nongeostationary systems;
5. Part of the new satellite-mobile services, either geostationary or non-geostationary.

There are no doubt other strategies worth considering, and newer ones will emerge as the related technologies emerge. Out of this list, option 4 has been around the longest but has not been implemented so far. Early attempts, such as Teledesic, were too grandiose, while the entrepreneurs behind the scaled-down version seem to have lost the nerve to fully invest in the approach. The first three options are all the subject of one or more ventures. Option 3 has both transparent and processing versions coming onstream in 2004–2005, and the market will no doubt give its verdict.

If one of these options is the basis for the business strategy proposed in the business plan, a couple of others could also be briefly discussed, along with the reasons that they were dropped. The rationale can be purely financial or a combination of market size and its credibility. It is important that the final choice is seen as a result of a dispassionate process of selection of the best strategy rather than being interpreted as the personal choice or preference of the management team.

System Infrastructure

In the initial stages of a satellite-based project, almost all of the capital is devoted toward building the infrastructure. Major elements are the satellites with their monitoring networks, uplink and content facilities, and if applicable content generation and processing facilities. For a new enterprise, the final system architecture may still be under development and a topic of internal debate at the time of the business plan development for the early funding campaigns. This by itself is not necessarily a negative, provided the plan and the presentations convey the confidence and ability to make the right and dispassionate choices on the part of the teams already onboard. Such selection processes can be quite time consuming and may require informal involvement of prospective vendors as well.

The plan document, as a minimum, should present the likely architectural scenarios and the *business-driven* criteria the management team plans to apply in order to select the final architecture. For each of the scenarios presented, there should be credible cost and schedule estimates. These schedules should include the time needed for the objective down-selection process itself. Last, but certainly not least, the development costs and schedules for the consumer equipment for each architectural scenario should be presented.

Products and Services

With the business strategy clearly identified and justified, the plan should then capture the different elements involved in the actual conduct and operations of the proposed business.

The discussion on business strategy should logically lead to a clear enunciation of the planned products and services of the company. The focus here should be entirely on the users of the system, whether the users are other companies or a large number of consumers. As far as possible, clear definitions of the service should be provided together with their roll-out schedule. If it is a consumer service, this section should identify either existing products or a summary of the arrangements in place for their development, production, and distribution, either through in-house efforts or through identified contractors and distributors. A cost center often ignored or understated in plans is the cost of distribution and collection of service revenues.

Operation

A top-level scheme should be provided in the plan to show how the planned services and products will be provided in the designated geographic areas. This section should include some details of actual operations, such as service roll outs, customer service centers, and mechanisms for revenue collection. It should also address the staffing and associated costs of all facilities, including those necessary for operating the system infrastructure.

Company Management and Advisors

This section in many cases determines the final outcome of investment-related efforts by a company. For a startup, the plan and its presentation are often a direct reflection of the management team. While external consultants and advisors are obviously necessary and even desirable for covering all aspects of the project, in the final analysis the investors look for talent and depth in the management teams in terms of leadership, business-acumen, decision-making, and team-building qualities. More often than not, a management team with a balanced set of capabilities, rather than just brilliant achievements in few narrow aspects, is seen as most likely to succeed and to sustain the project through the inevitable lean times.

This section should be supported with short and clear resumes in the appendixes.

Risks

A clear and frank discussion of all types of risks is almost a requirement of modern business plans. More and more, this section has begun to take the shape of almost a legal requirement as a safeguard against possible lawsuits by disappointed investors. However, that need not be the case in all cases, unless specified as requirements by national regulatory authorities.

Notwithstanding such possible legal requirements, it is very desirable to capture in the plan some of the relevant risk factors discussed in Chapter 4. As the satellite industry moves more and more toward consumer services, the time-critical market shares can be quite adversely affected by launch failures. In some cases, some kind of business insurance in addition to the satellite/launch insurance itself may be insisted upon by perspective investors.

Financials

This section should bring together all costs and revenues in a rigorous manner, as dictated by prevailing financial and accounting standards and practices. While the details should be relegated to the appendixes, the body of the report should include all of the key elements in clear and understandable language.

Readers of this section would generally expect full cognizance of some of the recent pitfalls in the satellite industry, as discussed in earlier chapters. Most of the investors will expect to see the impact on cash flow positive time frames and returns on investment of such factors as delays in availability and distribution of consumer equipment, delays in satellite programs, and launch failures.

Summary

This chapter has provided general guidelines for preparing a business plan as part of an IBSP. A business plan can meet multiple objectives; however, the most common and difficult objective is obviously raising substantial funds for satellite-based projects. Due to the critical role played by spectrum and orbital locations (akin to real estate in business districts), satellite-based businesses in the past have given relatively less importance to other traditional business aspects, relying quite successfully on real barriers to entry to competitors without these assets. However, recent experiences and the growing role of consumer-based services in the satellite sector require that business plan documents treat other aspects with equal professional treatment.

Reference

[1]   Truitt, W. B., *Business Planning,* Westport, CT: Quorum Books, 2002.

6

System Planning

Having looked at the development of business strategy and business planning in the previous two chapters, we now move to another important activity in the IBSP, system planning (see Figure 3.1). For the ultimate success of the IBSP, an efficient system-planning group is a central function that acts as a binding force for multiple functions almost throughout the process from vision to final operations. The system-planning function fulfills this responsibility by acquiring adequate expertise *and jargon knowledge* to provide support to and interface with practically all other functions.

In spite of frequent vindications about the value of system planning, the fact remains that the concept of *planning* is not universally recognized or accepted, and there continues to be a wide diversity of opinions about this function. Part of this reaction is just a result of semantics. Often "planning" is interpreted as one study leading to another without a definitive outcome. Some of the skepticism about planning as a concept perhaps has its origin in the Cold War days shortly after World War II. For several decades, a planned economy was viewed primarily as a creation of the socialist and communist philosophies espoused by the former Soviet Union. Planning and plans were seen as mere tools to make the politicians look good by promising Utopia to the masses, unfortunately always in the distant future, at the end of the never-ending rainbow of planning cycles.

At the other end of the spectrum, several examples vindicate the importance of planning. The following summary for the recent cataclysmic epoch in the history of the telecommunication world provides a concrete example of the importance of what we will call *system planning* and is a dramatic example of what happens when a decades-long system-planning culture is suddenly eliminated.

The almost never-ending explanations for the recent precipitous decline of the telecommunication industry rarely include the impact of the landmark divestiture of the Bell System in 1984 by Judge Harold Greene [1]. What is sometimes included is the more recent Telecommunication Deregulation Act of 1996 [2]. The underlying rationale for both of these historic decisions was to transition the earlier monopoly in services and manufacturing of telecommunication products to a more open environment with the objective of providing the consumers the intrinsic benefits of competition. While these objectives have been met to an extent, there was a colossal meltdown of the telecommunication sector around the turn of the century. As the successive corporate disasters unfolded, a pattern began to emerge regarding the underlying reasons. The most common reasons identified so far are mismanagement, greed, and overcapacity. The first two are complex factors and are not necessarily unique to telecommunications. To understand the third factor, overcapacity, particularly for optical fiber cables, it is helpful to look briefly at the history.

Prior to its breakup, AT&T was a huge monolithic organization, with the Bell system vertically integrated over the entire value chain, including the world-famous Bell Laboratories, the giant Western Electric, and a group of regional "Bell" operating units providing universal telephone services conforming to centrally specified standards. All local and long-distance network expansions were carefully *planned* and coordinated within each region and across the country, as well as for links to the rest of the world, with active monitoring of the related capital investments by the designated federal and state regulators. While it was generally accepted that the Bell system was arguably the most cost-efficient phone system in the developed world, it was nevertheless a monopoly—and hence an automatic anathema to those wedded to the principles of free and competitive markets. The "new" competitive regimes that followed the breakup did provide some of the anticipated benefits; however, in this process they also weakened the long-established binding mechanisms for coordinated infrastructure investments. For large and interconnected systems spread over huge geographical areas, such weaknesses take years to surface. Meanwhile, a host of old and new companies made a large number of apparently uncoordinated—and in hindsight reckless—investments right in the middle of the *economic bubble* days at the end of the last century. The scale of these telecommunication investments was egged on by irrationally exuberant prognostications about the benefits of broadband [3] on one hand and the gilded investment gurus on the other. As a result, the telecommunication sector was no longer just the victim but became in fact one of the prime movers for the economic bubble sweeping through the economy. The segment of telecommunications that witnessed the highest impact was the optical fiber industry, which created at a record pace thousands miles of excess capacity of optical fiber pathways not terminated anywhere and hence totally useless. Ironically, this huge installation of new fiber cables took place precisely when the phenomenal technological advances of wavelength multiplexing were becoming available for multiplying severalfold the capacity of existing fiber cable routes. That this investment binge led to some of the largest bankruptcies was almost preordained by the sudden elimination of planning at the national level.

The adverse impact of this telecommunication bubble, several hundreds of billions of dollars and still counting, was not just confined to optical cables but was felt up and down the complete value chains. The satellite industry was no exception, as telecommunication entities are major users of satellite systems and in many cases also major investors.

What Is System Planning?

In modern times, the underlying concepts of planning are indeed appreciated, but very often under different nomenclatures under different environments. For example, modern management concepts utilizing tools such as PERT/CPM and Gantt charts are currently an essential part of almost all projects, except for the very simple ones. For a large and complex program, a significant part of what is carried out by the project manager through her office staff is in fact what could be considered system planning. In general, what is called project management in one situation may well be called system planning or system engineering in another context.

Within the satellite sector itself, one can also identify different definitions for such a function. The larger international operating entities did start with the system planning nomenclature, but there is a recent trend to imbed this activity within business development, perhaps to ensure a greater cohesion with the strategic evolution of the company. Sometimes, this function is labeled system engineering and made part of the engineering division. On the manufacturing side, it is relatively rare to see the use of the system-planning terminology itself; rather, the most common nomenclature is system engineering, closely followed by business development and project team/office.

The best way to capture the scope and definition of system planning is to look at its major activities, as shown in Figure 6.1. The chart also shows typical interactions and interdependencies between system planning and other functions. We will now address some of the major activities of system planning with particular emphasis on satellite-based systems.

System Architecture Selection and Development

Selection of an appropriate architecture responsive to the mission and business strategy of an enterprise can be an important contribution of system planning in

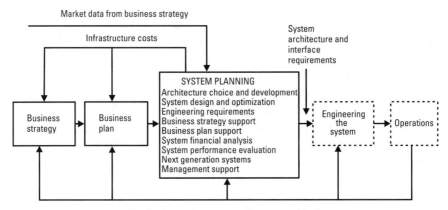

Figure 6.1 System planning: functions and interactions.

terms of enhancing the financial viability of the project and the ultimate customer satisfaction.

The processes of developing alternative system architectures and downselecting to an optimum architecture must take into account several environmental factors that were discussed in Chapter 4 as part of the business strategy development process. Professors de Weck and Crawley [4] have elegantly combined factors likely to be encountered in satellite-based systems and related technologies in one chart reproduced in Figure 6.2. They also provide some useful general definitions on why products succeed and why they fail [4]:

- A product is judged to be *successful* if it "delivers external function to the operand which the beneficiary perceives as having competitive value, while meeting enterprise objectives (value to shareholders) and conforming with pseudo-regulations (value to society)."

- Along the same lines, products are judged to have *failed* if they "do not deliver extended function to the operand ... with value at price, do not meet enterprise objectives, OR other products have emerged which deliver more value at price."

These definitions, while a bit terse on first reading, are quite instructive and underscore that successful products are invariably a result of a good and well-thought-out system architecture, which in turn is responsive to the mission that started the enterprise. Professor de Weck and his associates have also applied these system architecture trade concepts to satellite-based systems [5, 6]. The methodology adopted is sufficiently rigorous and generic. The approach

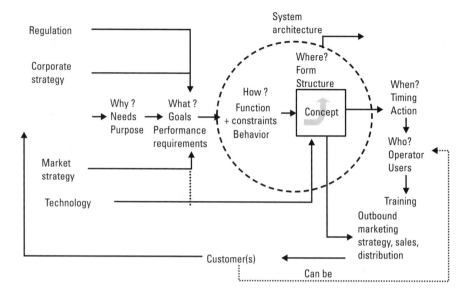

Figure 6.2 System trade framework. (*From:* [4]. © 2001 Olivier L. de Weck. Reprinted with permission.)

and results of this work as reported in [5] are summarized in Case History I at the end of this chapter.

System Architecture Examples

Satellite-based systems almost by definition involve large geographic areas, often across large bodies of water and uninhabited terrains. It is not surprising, therefore, that proper system architecture often plays a crucial role in meeting the mission objectives. A few examples, some of them classic historic milestones, are recalled here:

1. *Geostationary orbit.* Arthur C. Clarke came up with the now well-known three-satellite geostationary network as early as 1945, long before any satellites were built or even their predecessor, microwave systems, existed [7]. His fundamental concept of three equally spaced radio transmitters at a height of 36,000 km, each appearing stationary as seen from one-third of the Earth below, is an elegant, simple, and almost timeless example of optimum architecture development.

2. *Colocated spacecraft.* Once the geostationary satellites were proven in the early 1960s, they soon became natural nodes or *switchboards in the sky* providing interconnectivity to a growing number of ground antennas, often serving multiple nations on both sides of oceans or large land masses. The increasing traffic required progressively higher

capacity from such satellites. Often these capacity demands would exceed what either the available spacecraft platforms or rockets could accommodate. Development of new spacecraft platforms and launch capabilities is quite expensive and time consuming. This led to an alternative architecture, wherein more than one spacecraft would be spaced sufficiently close in the geostationary orbit to simulate one large satellite as seen by antennas of appropriate beamwidths on the ground. While such an architecture led to the duplication of some of the sub-systems, such as satellite antennas, it was often more efficient both in overall costs and time. Operators who have adopted this type of architecture often stayed with it even when larger spacecraft and launchers became available, as it provided two important benefits: staggering of investments and an opportunity to better tailor the system capabilities to the changing markets and technology through successive launches. In other words, such a *rolling* architecture enabled the operators to better manage financial and market risks when compared to one large spacecraft. The most successful examples of this multiple-satellite architecture are the SES and Eutelsat clusters, mainly for television distribution and broadcasting, as briefly described in Chapter 2.

3. *National Aeronautic and Space Administration Tracking Data and Relay Satellite Network.* For the historic Apollo missions to the Moon in the 1960s, the National Aeronautic and Space Administration (NASA) used the INTELSAT three-ocean network, closely following the architecture envisioned by Clarke two decades earlier. These satellites provided continuous access to the astronauts but required extensive earth station installations around the world in order to bring real-time information to the NASA headquarters. As the planning started for the later missions, with the shuttle as an integral element, a radically different data relay and management architecture was conceived to more efficiently match the needs of all of the LEO facilities, such as the shuttle, the International Space Station, and other government applications in similar orbits. For the real-time transport and management of a voluminous amount of narrow and wideband data to these LEO systems, NASA developed an efficient "upside-down" architecture, consisting of just one large earth station in New Mexico and several relay satellites in geostationary orbit. Figure 6.3 shows the basic concepts underlying their Tracking Data and Relay Satellite (TDRS) System [8].

    This architecture was fully responsive to NASA's needs and had several novel features. All communications from platforms anywhere in the LEOs are received by one or more of the TDRS satellites and then sent down to New Mexico facilities. A lesser known feature of this architecture is to shift the complexity of the antenna

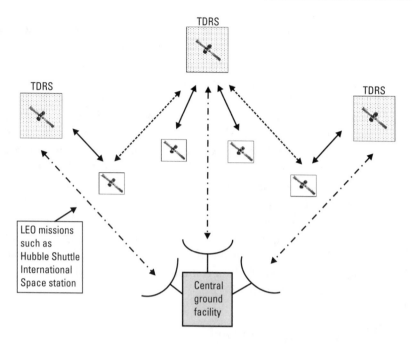

Figure 6.3 TDRS network concept.

beam-forming network to the ground. While this required a much larger feeder-link bandwidth, it substantially reduced the complexity and power demands onboard the spacecraft. In addition, the beam-forming software could be much more easily upgraded with changing needs.

4. *LEO systems for mobile communications.* While both Iridium and Globalstar targeted the same market of land-mobile communications, they adopted different system architectures. Iridium chose polar orbits linked by intersatellite links, the whole constellation completely encircling the globe, providing peer-to-peer communication with little or no need for the existing terrestrial networks over land masses. On the other hand, Globalstar adopted inclined circular orbits, using the terrestrial network whenever feasible. This architecture reduced the total number of satellites and eliminated the need for intersatellite links. As a result, the total investment for Globalstar was lower than that for Iridium for almost the same lifetime capacity. (See Case Study I for a quantitative comparison of these two systems.)

5. *Inclined elliptical orbits.* Soon after the geostationary orbits were demonstrated in the 1960s, it was recognized that while the geostationary architectures had the elegance of simplicity, they were not very

efficient for parts of the Earth away from the equator due to low elevation angles and hence susceptibility to blockage and interference from terrestrial systems and extraneous noise. The system planners went back to the basic orbital mechanics, and a number of alternative architectures have continued to emerge for the last 40 years. The well-known Molniya system launched by the Soviet Union in 1965 is active to this day and provides an elegant solution for its coverage areas in terms of higher elevation angles and repeatability of ground tracks. To serve mobile customers, the 24-hour Tundra orbit developed toward the late 1980s provided an elegant solution that has recently been adopted by the Sirius digital radio system in the United States [9, 10]. Several other system architectures and concepts have extended this approach to higher frequencies. Recently, inclined orbits have been demonstrated to not only provide higher elevation angles, but to also allow severalfold reuse of the frequency spectrum away from the equator [11].

System Design and Optimization

In general, a well-prepared system architecture provides a reasonably complete scheme for responding to the identified marketing needs. Depending on its underlying complexity and scope, however, it may not be by itself quite ready for actual implementation. Furthermore, while first estimates of overall costs and schedules may be available, they may not be the most optimum. In many respects, a system architecture for a satellite system is somewhat akin to an architecture for a city or even part of a city. In either scenario, the next logical step is to conduct a more detailed system design and optimization, before concrete can be poured or metal cut in the factory.

Like many other interdisciplinary activities, there is often a controversy as to which group should be entrusted with the lead responsibility for the system-design phase. Generally, the competing groups are engineering and system planning (or equivalent). In a structure similar to the IBSP developed in this book, this activity is best led by the system-planning team in close cooperation with the engineering team. While system planning brings the knowledge and understanding of the overall business strategy and the targeted market share, a good engineering team provides up-to-date knowledge of the state of the art in all major areas of the planned system architecture. Close cooperation between these two teams can be expected to lead to the best system design while still remaining within the boundaries of what the management and the chosen business strategy wishes to realize in the marketplace. We will now address some of the major activities that take place during the system design and optimization phase of system planning.

Industrial Technology Assessment

Selection of technologies at the detailed design level is clearly the domain of the engineering and the industrial phases in each subsystem or module. However, right at the outset of the system-design process (preferably even earlier), it is important to select a set of broad technology directions in order to establish realistic and sound business-like boundaries. Such a top-level assessment of the technology readiness in the industry should be a cooperative exercise of the research and development specialists, system planning, and engineering teams and should be conducted through a well-constructed and face-to-face survey of industrial capabilities. This process should be quick and not a long-drawn one if it has to meaningfully influence the system-design process.

The key technologies should be selected on the basis of the business objectives and planned services for the system. Some of the key assessments that can benefit the system design of a satellite-based system in this decade are likely to include the following:

1. For direct broadcast systems, perhaps the most important assessment is the industrial viability of the planned consumer equipment in terms of its functionality and unit cost under bulk production. If it is a new item, it is also critical to get a credible assessment for the expected development costs and time frames. The achievable technical parameters of the mass-produced consumer equipment directly influence the design of the satellites, while the final cost to the consumer can determine the ultimate success of the overall business strategy.

2. Closely linked to item 1 and perhaps next in importance is an assessment of the most efficient source encoding, access, and broadcast techniques. While not generally well known, advances in source-encoding technologies for voice and video have played as important a role as high-power spacecraft in the realization of modern broadcast systems.

3. Competitive launch vehicle capabilities and their reliability records are also important.

4. Capabilities of competitive spacecraft buses or platforms in production should be included. For direct broadcast systems, the onboard power is generally a critical parameter, while for telecommunication and distribution systems, the maximum weight capability can be important.

5. For satellites systems requiring a high degree of frequency reuse (e.g., for mobile applications and for Ka-band multiple-beam systems), it can be important to know the largest diameter spacecraft antennas available in the industry.

6. With the growing use of onboard processing, it is also important to have an assessment of the processor capabilities in the industry in terms of communication throughput and power dissipation.

Market Demand Compatibility

With these technology assessments in hand, the process of converting the selected system architecture to a sound system design can start. An important step here is to substantiate in adequate detail the ability of the system to meet the demand projections over the project time frames. At the end of such a process, it is not that uncommon to find that the overall architecture itself may need some changes. The process of confirming the compatibility and match of the system design with the market requirements can vary depending on the targeted market and application.

1. For telecommunication services, the demand projections are generally provided in the form of connectivity matrixes between key earth stations, regions, or even countries covering the system time frames. In the system-design stage, these matrixes are translated into major communication parameters of the satellite payload, such as the number of standard transponders, up- and downlink coverages, and onboard connectivity matrixes, either static or dynamic. The types of antennas on the ground—existing and planned—influence the satellite mass and power requirements as well as the overall system capacity in terms of channels or megabits per second.

2. For one-way broadcast systems, the audience size in terms of number of receivers has no impact on the system design. On the other hand, the sizes of the coverage areas, the gain/temperature parameter (G/T) of the consumer equipment, and the required bandwidth in the satellite all have a direct impact on the satellite payload design.

3. Mobile satellite systems have all of the requirements of telecommunication satellites, but with a large number of *very small earth stations* in the form of mobile handsets. The system design of such systems can therefore be quite challenging in several respects, as has been experienced through several recent systems, such as Iridium, Globalstar, Thuraya, and ACeS. Unlike these first generation systems, the challenge for system designers in this decade is to develop satellite systems, whether LEO, medium-earth orbit (MEO), or geostationary earth orbit (GEO), compatible with today's small cellular handsets with limited power and size.

4. Broadband systems today generally connote two-way systems to provide high-speed Internet access to businesses and homes. In many

respects, they are as challenging as mobile systems, but with emphasis on higher user bandwidth rather than on the consumer equipment size and power. While the current terminals are fixed installations, the challenge in the not too distant future will be to provide broadband to mobile users in order to match the capabilities of next generation cellular and wireless systems. Experience of the very first such broadband systems demonstrated that these systems are not direct and simple extensions of broadcast systems. The satellite capacity requirements of large numbers of (return) uplinks have a significant impact on the overall system cost effectiveness.

Compliance with Regulatory Requirements

A new entrepreneur entering the satellite business quickly learns about the importance of regulatory compliance. Generally, the first such requirement is the allocation of the necessary orbital spectrum and license from the relevant national authorities. As the architecture development and system design proceed, at several stages compliance with regulations can influence the design approaches and often the system economics. A few examples follow:

1. License for a full or partial orbital location comes with several responsibilities typical of a good neighbor. There are strict guidelines for allowable interference to adjacent satellites, space debris, and stationkeeping. A new kid in the block has often a nonreciprocal obligation to not impact existing services on adjacent satellites. The spacecraft design has to meet ever-increasing constraints of space debris not only in its lifetime but also in its "afterlife" in terms of moving to slightly higher orbits to avoid collisions. The stationkeeping requirements are not a serious constraint with current technology, although collocation with several other satellites can be more complicated, particularly if multiple operators are sharing the same location.

2. Several frequency bands are shared between terrestrial and space users. This generally places an upper limit on the maximum flux density allowed from the satellite. This constraint can seriously impact the capacity of telecommunication satellites and the smallest antenna sizes for broadcast systems. As an example, the L-band allocation for digital radio is a shared one, while the S-band allocation for similar service in the United States is not.

3. Shared bands can also impact the design of earth-station antennas. These antennas are required to meet off-beam emission requirements. This can sometimes lead to a minimum size requirement for the user antennas.

Compliance with Financial Criteria

One of the major responsibilities of the system-planning group in the IBSP is to ensure that throughout the process, the system being developed and implemented will meet the financial criteria committed to the investors by the company's management. While the overall financial oversight responsibility of the company falls on the company chief financial officer (CFO) or equivalent, it is the system-planning group that can and should ensure that at all stages the decisions and choices being made are compatible with the financial targets and criteria. Often the seeds of cost or schedule overrun can be sown through a seemingly simple and innocuous decision at a working level in any of the IBSP activities. How can the system-planning group achieve this objective without being accused of being too overbearing or going outside its domain and stepping on the toes of the CFO's teams? The answer, as is generally the case in such situations, lies partially in being seen by all of the teams involved as a helpful partner presenting a system perspective coupled with the financial impact at all stages of the project. A few examples are:

1. System planning assists the business strategy and business planning activities through an approximate sizing of the required system and its costs to meet the targeted market share of services. This sets the financial targets in front of potential investors. This exercise may have to be repeated for each of the strategic options before the final business strategy is selected. It is worth emphasizing that the ability of system planning to fulfill this responsibility is a direct function of the past experience of its members coupled with their ability to gather outside advice as needed.

2. This process is repeated as the appropriate architecture is selected. It is quite likely that, by this time, there are changes in the external environment. These should be factored in and all other groups up and down the chain should be kept informed.

3. The system-design process is very likely to require changes in the financial numbers, as this is perhaps the first time that engineering experts and industry get involved in the process. The design process should be iterated until the financial results are close to what was committed to the investors. If not, the results should be shared before major commitments are made.

4. During the engineering phase, firm and definitive cost figures become available for all major items. As we discuss in the next chapter, system planning plays an integral role in the selection process in part through the financial evaluations.

Engineering Interface Requirements

Once the system design is firmed up, the main deliverable from system planning is the interface requirements for all major subsystems. These generally flow from the overall link design for the system. These requirements should be defined to allow relatively independent procurement of each major subsystem during the engineering phase. A test of the comprehensiveness of these requirements is that once all of the subsystems are in place, the overall end-to-end performance of the system design will be met. The next chapter will elaborate on some of these requirements.

In summary, at the conclusion of the system design and optimization phase, the project should have met the following criteria:

- The system design is responsive to the market needs over the lifetime of the spacecraft with reasonable fill factors.

- Technologies called for are available in the industry with acceptable risk.

- The financial analyses confirm that the project can meet the criteria laid down by the management and the investors.

- The system is reasonably tolerant of the likely variances in demand and has a reasonable degree of fungibility and scalability of assets.

- The system is compatible with regulatory requirements.

- The system has clean and definable interfaces compatible with competitive procurements of different subsystems.

At the end of this chapter, Case History II documents a process of interactive system design for the INTELSAT VII program around the 1990s.

Other System-Planning Activities

The system-planning box in Figure 6.1 lists some of the other activities typical for this group. These can include:

- *Support for system performance evaluations.* An important example here is the assessment of the system and financial impact of any deviations requested by satellite and other contractors.

- *Next generation systems.* The strength of the system-planning team in terms of supporting several other groups derives in part from its ability to take a longer term view of the evolution of the company's business, both where it came from and where it could be going. Such groups can conduct valuable advance thinking and brainstorming for the next generation systems while keeping an eye on the ongoing implementation and operational activities.

- *Management support.* A good and responsive system-planning group can be a valuable information resource for top management. Furthermore, its breadth of coverage and involvement in multiple activities can make it an excellent sounding board for providing insight into "what if" scenarios being looked at by senior management. An internal evaluation of a possible acquisition or merger is a concrete example of such a role.

Case Histories

Case History I: Architecture Trade Methodology

Olivier L. de Weck and Darren D. Chang [5] have presented an architecture trade methodology that can not only provide quantitative confirmation if the market and technical assumptions are indeed correct, it can also reveal some new information about the system even before it is actually built. While the published methodology specifically addresses LEO personal communication systems, it can be applied to a wide range of satellite applications discussed in this book.

The analysis presented starts with the input-output mapping of the planned constellation model of the type shown in Figure 6.4. Five specific sets of vectors are the inputs to the model:

- The *design vectors* capture the key system parameters, such as the type of constellation, transmit power, network architecture, and lifetime. For each of these design vectors, the range over which it can be varied are also fed in.

- The *constant vectors* capture certain technical approaches that are often predetermined by the user equipment, for example.

- The *requirements vectors*, as the name implies, provide the key system performance, such as the bit error rate and the link margin.

- *Policy decisions* are the inputs from the external environment, typically regulatory mandates, both national and international.

- The model outputs are the *objective vectors*. For a typical satellite-based system, these include lifetime costs and capacity and costs per function.

In order to get the range of answers needed for the objective vectors, a series of computational modules are required. These can be generally

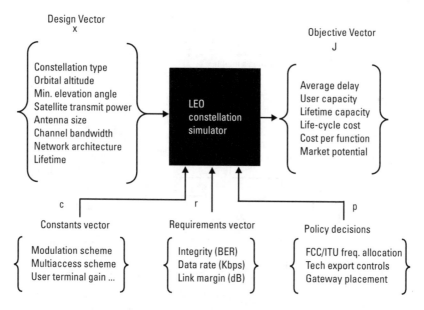

Figure 6.4 Input-output mapping of constellation model. (*From:* [5]. © 2002 Olivier L. de Weck. Reprinted with permission.)

categorized in three groups: technical, cost, and market, as shown in Figure 6.5. While the technical modules are largely standard for a broad group of applications, the cost and market modules need inputs based on the current and anticipated trends in the relevant areas.

Figure 6.6 shows the kind of trade results that can be obtained. The example shown is the one published by the authors for LEO systems. The trade space shown covers a wide range of system lifetime costs and capacities. The lowest capacity systems are mapped at the lower left, while the largest capacity systems feasible with the range of parameters assumed are at the top right. As a measure of the accuracy of such trade methodologies, the figure also shows the actual and simulated two LEO systems, Iridium and Globalstar.

Case History II: The INTELSAT VII Program

An Early Example of Integrated Approach for System Design

Before the INTELSAT VII program was initiated around 1986 [12], there had been six series of spacecraft, typically with multiple identical or similar spacecraft in each such series. (The first such series in the early 1960s was in fact the very first geostationary communication satellite ever built.) Each successive series was bigger than the previous one, in just about every manner. The spacecraft bus or platform was bigger, had more power generation capability, and higher communication capacity than the previous one.

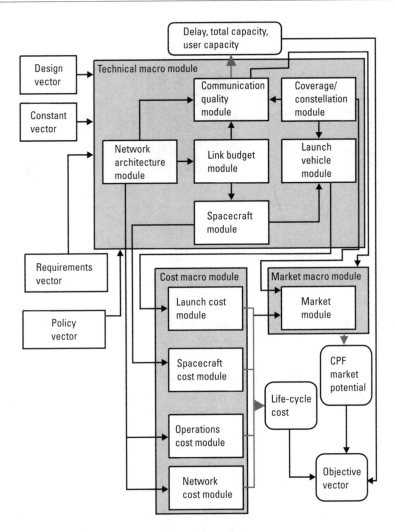

Figure 6.5 Constellation simulation block schematic. (*From:* [5]. © 2002 Olivier L. de Weck. Reprinted with permission.)

The INTELSAT VII series was different in several respects, not only in its mission or objectives but also in the overall management approach adopted for its planning as well as engineering. Unlike all of the previous series, the VII series was designed not for the mid-ocean connectivity roles in the Atlantic and the Pacific regions, but for orbital locations closer to the land masses in each of the three ocean regions, including the Pacific. In terms of communication capacity, for the first time the benchmark was not the largest antennas in the system (the standards A, B, and C), but the smaller antennas at both C- and Ku-bands. The diverse and different roles designated for this series placed sometimes conflicting demands on the basic design of the spacecraft payloads.

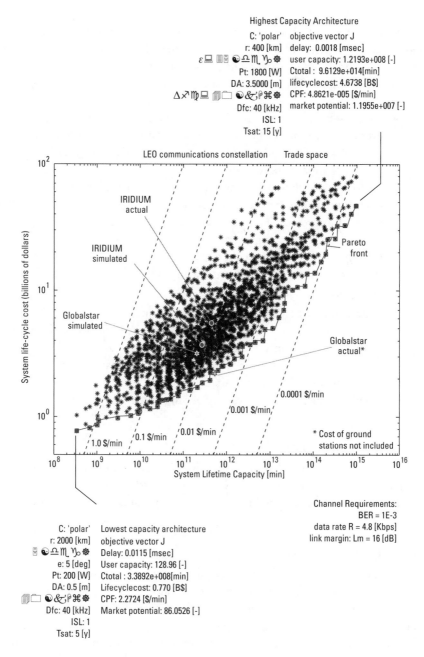

Highest Capacity Architecture

C: 'polar'	objective vector J
r: 400 [km]	delay: 0.0018 [msec]
ε⬜ 🖥🖥 ●⊒🜔 ℕ Yo ✵	user capacity: 1.2193e+008 [-]
Pt: 1800 [W]	Ctotal : 9.6129e+014[min]
DA: 3.5000 [m]	lifecyclecost: 4.6738 [B$]
Δ↗🝆⬜ 📑☐ ●&;𝒫⌘✵	CPF: 4.8621e-005 [$/min]
Dfc: 40 [kHz]	market potential: 1.1955e+007 [-]
ISL: 1	
Tsat: 15 [y]	

LEO communications constellation   Trade space

System life-cycle cost (billions of dollars)

IRIDIUM actual

IRIDIUM simulated

Globalstar simulated

Pareto front

Globalstar actual*

0.0001 $/min

0.001 $/min,

1.0 $/min   0.1 $/min   0.01 $/min

* Cost of ground stations not included

System Lifetime Capacity [min]

Channel Requirements:
BER = 1E-3
data rate R = 4.8 [Kbps]
link margin: Lm = 16 [dB]

C: 'polar'	Lowest capacity architecture
r: 2000 [km]	objective vector J
🖥 ●⊒ℕ Yo ✵	Delay: 0.0115 [msec]
e: 5 [deg]	User capacity: 128.96 [-]
Pt: 200 [W]	Ctotal : 3.3892e+008[min]
DA: 0.5 [m]	Lifecyclecost: 0.770 [B$]
📑☐ ●&;𝒫⌘✵	CPF: 2.2724 [$/min]
Dfc: 40 [kHz]	Market potential: 86.0526 [-]
ISL: 1	
Tsat: 5 [y]	

Figure 6.6 LEO constellation trade space. (*From:* [5]. © 2002 Olivier L. de Weck. Reprinted with permission.)

In order to achieve the objectives outlined earlier, the overall planning and engineering approaches adopted broke several new grounds, compared to all of the previous series. In some respects, they provided an early confirmation and vindicated the benefits of an integrated approach, expanded in this book to cover an even broader range of activities and functions.

Major examples of this approach are:

- The system planning function was much more tightly coupled than ever before to the marketing and operational functions on one side and with engineering and finance on the other. Tangible benefits of this approach were realized right from the beginning and throughout the project. The alternative architectures were tested directly with the operations teams before being subjected to any detailed engineering effort, thus providing the ultimate users an early and positive say in the whole process. The engineering teams were collocated with the system-planning team, thus providing each team validation of its assumptions throughout the process of narrowing down the architectural options and spacecraft sizing.

- A decision was taken to resist the tradition of designing a bigger and better platform for the spacecraft and to instead utilize available platforms. This not only reduced the overall costs and schedules, it also acted as a welcome brake on adding nice-to-have functions to the requirements. (The spacecraft unfortunately did not get delivered on schedule, largely due to a variety of programmatic issues.)

Benefits of Integrated Approach

The close integration of planning and engineering, coupled with the internal discipline to keep the spacecraft size within the industry capabilities, was instrumental in several innovations in the overall design approach. A few examples follow.

Antenna Coverage Flexibility

The demands in terms of coverage requirements from the designated orbital locations in the three-ocean region required a fairly high degree of flexibility in the antenna system. However, the system-planning and engineering teams took up the challenge of developing a simpler and hopefully better approach than the previous series, wherein large, heavy, and complex switching networks were used to provide interocean operability. Even with more modern beam-forming technologies, the antenna design would have been unduly complex.

After considerable amount of interaction and analysis, the teams came out with the innovative approach of exploiting the very geography of the planet, which was initially presenting this daunting problem. Specifically, by inverting the spacecraft at certain locations, all of the requirements could be satisfied, without any additional complexity over and above what would have been required by a much smaller set of locations. Figure 6.7 shows the results. As can be seen, while the beam shapes at the two locations are identical, the Earth underneath is inverted. This was achieved in orbit by a simple one-time command to the spacecraft!

Sharing of Capacity Across the Oceans

While relatively straightforward as implemented, this innovation gave the benefits of a six-times reuse antenna without incurring the cost of a full five-times reuse payload. As Figure 6.8 shows, by simple switching of the transponders diagonally, the capacity can be deployed where needed.

C-Band Spot Coverage

In all previous series before *INTELSAT VII*, the C-band capacity was provided either through large hemi/zone beams covering the Eastern and Western hemispheres or through a few global beam transponders. Demand patterns indicated a third category: spot beams at C-band similar to those at Ku-band. However, permanent allocation of spectrum to such beams would have led to reduction of overall capacity. An innovative solution that emerged was to add a new C-band spot antenna, which would borrow one or more of the global transponders as needed and beam a much higher power beam to areas of interest.

In summary, the *INTELSAT VII* program tested some of the early concepts of integration through active communication and feedback between different functions in a complex program. Not only did this approach substantially

Normal orientation

Inverted orientation

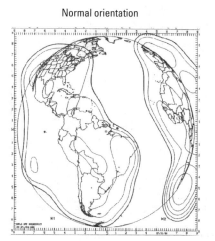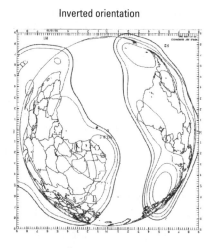

Figure 6.7 Example of coverage flexibility through spacecraft inversion in orbit. (*From:* [12]. © 1990 IEEE. Reprinted with permission.)

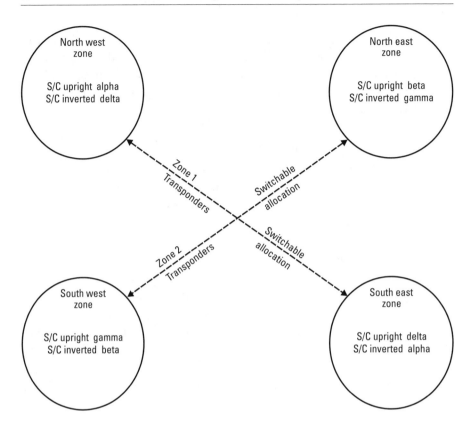

Figure 6.8 C-band zone flexibility. (*From:* [12]. © 1990 IEEE. Reprinted with permission.)

shorten the precontract duration compared to earlier similar programs, it also led to a very innovative and highly flexible design. In terms of the lessons of Chapter 2, the *INTELSAT VII* was easily the most fungible design of its time.

References

[1]   "The AT&T Breakup—One Year Later," *Wall Street Journal*, December 17, 1984.

[2]   "Telecommunication Act of 1996," http://www.fcc.gov/telecom.html.

[3]   Rivilin, G., *Wired*, Vol. 10, No. 7, July 2002.

[4]   de Weck, O. L., and E. Crawley, "System Architecture Trade Studies," Lecture Notes 16.882/ESD Engineering Systems Dept., MIT, Cambridge, MA, November 26, 2001.

[5]   de Weck, O. L., and D. Chang, "Architecture Trade Methodology for LEO Personal Communication Systems," *20th AIAA International Communication Satellite Systems Conference*, May 2002, paper 2002-1866.

[6] de Weck, O. L., et al., "A Parametric Communications Spacecraft Model for Conceptual Design Trade Studies," *21st AIAA International Communication Satellite Systems Conference*, paper 2003-2310, 2003.

[7] Clarke, A. C., "Extra Terrestrial Relays," *Wireless World*, October 1945, pp. 305–308.

[8] http://nmsp.gsfc.nasa.gov/range/rstim.pdf.

[9] Ashton, C. J., "Archimedes-Land Mobile Communications for Highly Inclined Satellite Orbits," *IEE Conference on Satellite Mobile Communications*, Brighton, England, September 1988, pp. 133–137.

[10] Briskman, R., and R. Prevaux, "S-DARDS Broadcast from Inclined Elliptical Orbits," *52nd International Astronautical Congress*, Toulouse, France, October 1–5, 2001.

[11] Draim, J., et al., "Beyond GEO—Using Elliptical Orbit Constellations to Multiply the Space Real Estate," *52nd International Astronautical Congress*, Toulouse, France, October 1–5, 2001.

[12] Sachdev, D. K., et al., "INTELSAT VII: A Flexible Spacecraft for the 1990s and Beyond," *Proc. IEEE*, Vol. 78, No. 7, July 1990.

7

Engineering the System

Finally to the nuts and bolts!

In the preceding chapters, we have made progress through several stages of the IBSP of Figure 3.1. We started with the company's mission, developing the general principles for analyzing the applicable external and internal environmental factors on the way to evolving an appropriate business strategy for the company. A well-developed business strategy was the starting point of two somewhat-interrelated activities, business plan and system planning, in Chapters 5 and 6, respectively. We will assume in this chapter that a responsive business plan led to the availability of the funds necessary for building the system in its entirety, including operational systems, and that the system planning activities described in the preceding chapter have successfully selected an optimum system architecture that in turn has been subjected to a rigorous process of design optimization. As we embark on the engineering phase, the preceding activities would have provided us with the end-to-end performance objectives for the system as well as the interface requirements for its major components.

Following the approach adopted in previous chapters, Figure 7.1 expands the engineering phase of the IBSP and lists some of its common activities. The engineering team receives most of its inputs from the system-planning team in the form of interface functional and performance requirements. During this phase, as the development and engineering work proceeds on major infrastructure elements, continuous feedback is provided to all other activities—preceding as well as following—on the actual costs and the level of realizable performances with acceptable risks.

The activities listed inside the engineering box in Figure 7.1 are typical for satellite-based systems. Obviously, their role and relative importance can vary

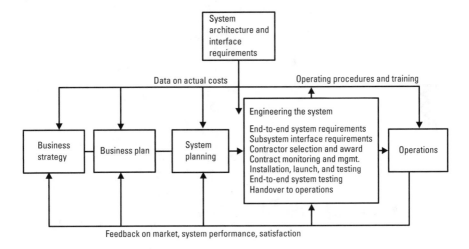

Figure 7.1 Engineering the system.

with different components of the overall system architecture, as we shall see later in this chapter.

Specification Creep

In spite of all the safeguards put in processes like the IBSP, one of the common pitfalls in many engineering projects of reasonable complexity is that once the engineering teams are assured of the funds, they tend to go full-steam ahead to design the most sophisticated and advanced system, often incorporating technologies and gizmos not because they are all really needed but largely because the engineers involved might have waited for years to deploy such features in an operational system. Often, the requirements at various levels get "stretched and twisted" to accommodate the so-called technologies of the future. In other cases, as the project proceeds, what started as an optional feature gradually morphs itself into a "must-have" requirement. This *specification* or *scope creep* is well known to experienced managers in practically all industries and can manifest itself at different stages of the program. For example, right at the outset, the interpretation of the broad system-planning requirements can be "revised" to accommodate new technologies. Or, the manufacturers and suppliers can try to swerve the requirements to specific technologies they might have already invested in order to gain a competitive edge in the bid-evaluation process. However, the most common culprits in this regard are often the "technical" experts of the customer itself, particularly those without direct fiscal responsibilities. As they walk around the corridors of the manufacturers' research labs and facilities, these experts become fascinated by a little feature here and a little enhancement there, and soon the management is looking at a major revision of the requirements. Such revisions are generally quite expensive, particularly if they emerge after the competitive process has been completed, often exposing the project to higher prices in a sole-source environment.

This may sound like an unnecessarily pessimistic assessment and perhaps too harsh a generalization. That may indeed be the case. In general, the interactive processes like the IBSP provide a self-regulating mechanism wherein the engineering teams get accustomed from the beginning to design and build systems that are driven by market requirements and committed financial returns. Furthermore, as discussed in the previous chapter, one effective way of avoiding any technology-driven specification creep is to make the system-planning team an active keeper of the requirements underlying the engineering phase, as shown in Figure 7.1.

Measures of Success for the Engineering Phase

Within the IBSP, it is also quite helpful to clearly establish early enough in the engineering phase what could be called *measures of success*. While these apply to the whole IBSP, they are particularly critical during the engineering phase, as that is when most of the large and irreversible financial commitments are made. For modern satellite-based projects, a few such measures will now be developed. Some of these can be seen as derivatives of the lessons described in Chapter 2.

Consumer Costs

The critical importance of meeting the target consumer equipment price and performance has been underscored several times in the preceding chapters. In almost all consumer fields, this is one of the important measures of success in the marketplace. For capital-intensive satellite systems, with a substantial debt load upfront, failure to meet the targeted *total* consumer costs sends a message to the investors that the forecasted revenues won't materialize, and the risk assessments and associated ratings of the company worsen fairly rapidly. Such as assessment generally assumes that there is a significant demand elasticity based on the consumer's overall costs, and therefore the demand could drop sharply beyond a certain price threshold.

Given the more than 15-year lifetime of modern spacecraft, the consumer-equipment design and configuration can go through several versions or even generations consistent with services' evolution and advances in basic electronic components. This indirect benefit of Moore's Law can often ameliorate the temporary adverse impact of the higher costs of the first generation consumer units. However, if the dominant portion of the consumer costs is system related (e.g., subscription costs), that can be more serious because it reflects the already sunk cost of the infrastructure and its ongoing operating expenses.

It is in this context that it is absolutely critical to confirm that the total consumer costs targets will be met before irrevocable and irreversible commitments are made on the spacecraft design and overall system costs, as discussed in more detail later in this chapter. This is indeed the most critical measure of success for consumer-oriented satellite systems. Within the IBSP, such an assessment should be made as part of business strategy and system planning, and then revisited before entering into the contractual commitments of every major engineering phase.

Schedule Compliance

Next to consumer costs, overall schedule compliance comes as a close second. There are two distinct aspects here. It is well known that schedule slippages in spacecraft programs generally mean higher costs, directly and indirectly. Depending on the contract terms, at least a part of such costs gets passed on to the customer (particularly if the contractor can claim that the customer-driven specification creep during the program was the culprit). The spacecraft cost increases can have a domino effect through associated spacecraft launch contracts and service-related marketing campaigns.

The lesser known but often more damaging aspect of this measure of success is the impact a schedule slippage can have on the ability to meet the forecasted revenues. For services targeted at consumers, the market forecasts can be heavily time critical, and serious schedule slippages can be quite devastating to the enterprise, particularly if there are emerging alternative choices also vying for the same or similar requirements. The two well-known examples are mobile services and broadband. As we have already noted, delays in the first generation mobile satellite services allowed the terrestrial cellular services to develop sufficiently to substantially raise the barriers to entry for the satellite systems. In broadband, this situation is still unfolding in different scenarios. In the United States, cable systems are capturing a major share of the market, while the digital subscriber line (DSL) systems are still constrained by technology and regulatory decisions. Truly competitive satellite alternatives, Spaceway and Wildblue, are scheduled to be deployed in 2005.

Overall, schedule compliance is an important measure of success for satellite-based systems, not only in terms of consumer and investor confidence but also in creating a better overall business environment for satellite-based systems.

System Flexibility and Fungibility

An ideal system architecture, of course, is one that can cater to multiple market segments and a wide range of market changes and vagaries. An every day example would be a multilane divided highway that can accommodate practically any kind of vehicular traffic, although a haphazard mix of speeds and vehicle size can substantially reduce the overall highway efficiency. A land-based wideband fiber route comes close, particularly if the repeaters can be periodically upgraded for performance and bandwidth. In the satellite field, standardized transparent transponders are a classic and successful example of flexibility and fungibility. Today a large number of such older transponders are providing advanced digital services that did not even exist on the drawing board when the satellites hosting these transponders were designed. They also are reasonably immune to changing demand patterns in a region by providing services through relatively wide beams, such as the hemispheric beams for international satellites and Continental United States (CONUS) beams and regional beams for Europe.

With the advent of newer services directly to the end consumers, extensive flexibility and fungibility has often been difficult to maintain. This is due to several reasons, including:

- Direct broadcast systems operate in unique bands often dedicated to just one service segment (e.g., television or radio).

- Mobile systems operate in relatively narrow band segments, and the required in-orbit capacity levels mandate a large number of spot beams and often custom onboard processors, all dedicated to just one service segment often using proprietary access techniques.

- Broadband systems operating at Ka-band also resort to high orders of frequency reuse coupled in some cases with onboard processing, once again reducing dramatically the fungibility of the expensive assets.

In summary, the engineering and planning teams have to continually weigh the system and marketing benefits of customization of the architecture on one hand and the flexibility and fungibility of the final system on the other. How they balance the two opposing attributes is another measure of their success.

Consistency with Investment Levels in Business Plan

Finally, the ultimate measure of success is the bottom line. At the start of operations, while the planned revenue streams are still ahead, bulk of the investments in the business plan are already committed or spent. The final measure of success is if the overall system meets the performance, cost, and schedule targets promised to the investors and the customers.

Types of Satellite-Based Infrastructures

We will now address how engineering infrastructures, responsive to these measures of success, can be developed consistently with the mission and business strategy of the company. In order to avoid repetitive treatments, we will first take a brief overview of the types of satellite-based systems currently in use and those likely to be implemented in the near future. In this overview we will expand on the generic structures discussed in Chapter 1, specifically the structures for one-way and two-way systems (Figures 1.1 and 1.2). From this overview, we will select major common items for a more detailed discussion, emphasizing the management and programmatic aspects of their engineering and implementation.

Telecommunication Systems

These are two-way systems by definition, although certain one-way services such as fax and data may also be provided. As is well known, the operational use of satellite technology started with such applications in the 1960s. The initial emphasis was for international links and soon domestic telecommunication applications also became popular. In all such applications, the interface at both ends was to national telecommunication networks, thus popularizing the name *gateway* (and more recently *teleports*) for the related earth-station complexes. With the rapid expansion of fiber cable networks in national networks, domestic and regional use of the satellite medium has declined considerably. However, international applications still hold a good portion of this market segment largely due to two unique attributes of the satellite medium: connectivity in orbit (the so-called switchboard in the sky) and the ability to reach far-flung areas at no extra cost.

The second group of telecommunication applications still demonstrating a steady growth is the one providing service directly to the ultimate user without going through the telecommunication network. These include the well-known very small aperture terminal (VSAT) networks for private corporate networks and thin-route applications in developing countries. Figure 7.2(a) expands on Figure 1.1 to show both modes of telecommunication satellites: gateway and VSAT modes. Because the gateway mode is generally the dominant user of the satellite resources, it tends to be the determining factor in the design of the overall system and the spacecraft. This can in some instances be to the disadvantage of VSAT-type network, which operates in the *power-limited mode* as compared with the bandwidth-limited mode for the former category.

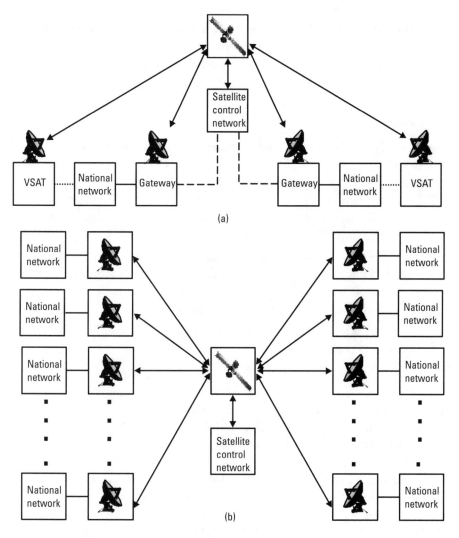

Figure 7.2 (a) Telecom networks, and (b) connectivity satellites.

Figure 7.2(b) is another adaptation of Figure 1.1 to show the connectivity mode or the switchboard-in the-sky mode. This can take a variety of forms, but for the traditional links between telecommunication networks, such satellites generally occupy mid-oceanic roles providing simultaneous visibility and connectivity to gateway-type earth stations in a large number of countries. It is this unique connectivity mode that has prevented undersea fiber cables from making as deep inroads in the international traffic as they have indeed done in national networks. Organizations like Intelsat, Panamsat, and SES continue to operate such satellites.

With the advent of the Internet, the nature of global telecommunication networks has changed and so have telecommunication satellites. They have had to quickly adapt to the unbalanced nature of such traffic and its operating protocols.

Broadcast Systems

Satellite broadcasting has evolved over the years in the following three broad categories:

- *Contribution.* The major components here are satellite news gathering (SNG), backhauling of special events, including sports, and exchange of programming between studios.
- *Distribution.* This category includes beaming of radio and television programs to local broadcasters and other affiliates, cable head ends, and hubs of DTU systems. This business sector currently accounts for a majority of transponders over land masses.
- *DTU.* This is a growing segment via high-power satellites and other enabling technologies.

Figure 7.3 has been adapted from Figure 1.2(c) to identify in a little more detail the basic building blocks of distribution systems. It shows two popular modes for television program distribution: local VHF/UHF transmitters and cable television systems. In the early stages, such systems were built using the same transponders used for telecommunication. In fact, the two service groups had just about equal influence in the definition of such transponders. To some extent such definitions are still in use even though a significant proportion of

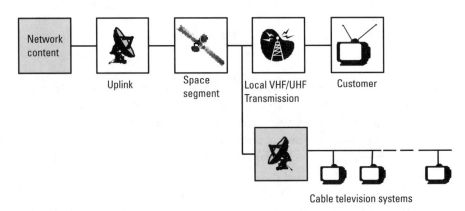

Cable television systems

Figure 7.3 Program distribution.

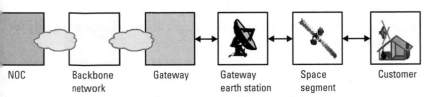

NOC | Backbone network | Gateway | Gateway earth station | Space segment | Customer

Figure 7.5 Broadband architecture.

differences as well as unique planning imperatives arise primarily from the fact that most links are now two-way instead of one-way. This has implications in terms of design and cost practically throughout the system. For example, the consumer equipment has to have a transmit capability as well; in fact, it is a small VSAT with much lower price ceilings. Equally significantly, the millions of uplinks from the consumers now make the spacecraft size and configuration sensitive to the audience size. Depending on the types of switched links planned, onboard processing often becomes necessary.

Chapter 2 noted the early attempts for broadband services by leveraging the Ku-band television broadcast systems. Based on this limited experience, at least three new systems will enter the critical broadband market by 2005.

Wildblue [1] in the United States and iPSTAR [2, 3] in Asia will continue the use of transparent repeaters for Internet protocol (IP) access and other services. The iPSTAR program is a particularly bold one, both in the size of market coverage and for the targeted capacity. Covering over 20 countries in Asia, it will have the ability to provide as much as 45-Gbps capacity through a single satellite. This level of capacity is planned to be achieved through 84 Ku-band spot beams and innovative coding techniques. Ka-band spot beams will be used for the gateways. What is particularly noteworthy is that such a high level of capacity is being achieved without onboard processing, thus preserving the fungibility of the system to a certain extent.

The third program, Spaceway [4], is for the United States and will achieve relatively modest capacity of 10 Gbps per satellite. However, what is unique about this satellite is that it will provide highly flexible peer-to-peer connectivity and much higher up and down transmission speeds that are typically needed by the highly attractive—and competitive—enterprise market. This will be achieved through a sophisticated onboard processor and hopping beam phase array Ka-band downlink antenna.

Mobile Systems

Mobile systems are almost as old as VSATs for telecommunications. Initially, they started only with the maritime market with simple bent-pipe L-band transponders either in stand-alone satellites or hosted on other mid-oceanic

satellites are being built exclusively for program distribution, parti
domestic markets.

Figure 7.4 shows a generic direct broadcast system, again
Figure 1.2. Broadcast systems are generally self-contained value ch:
content. They cater directly to millions of users through different l
mercial and distribution chains. Some of the content could be gen
the company, while some may be brought in recorded or in real ti
number of backhaul distribution satellites. Currently, the digital
create most of the content in house, whereas the video systems r
content providers outside the company. However, these modes c:
different parts of the world. The uplink facilities are responsible
content aggregation from different sources but also digital com|
encoding of all information channels. Radio broadcast systems als
trial repeaters, as discussed in more detail in Chapter 10 devoted to
of such systems.

Broadband Systems

This is a rapidly evolving service segment across several media. For m
large-enterprise customers, the predominant modes are via telecom
networks at T1 rates and higher. For small offices and home office
the dial-up access is rapidly getting replaced by faster techniques inc
grated services digital network (ISDN), DSL, cable modems, and in
satellite systems. For the satellite-based systems to have a significant
marketplace, they have to offer services competitive with DSL
modems in terms of quality as well as costs.

Figure 7.5 shows a generic architecture for a typical broadba
While at first glance such systems may resemble broadcast systems,

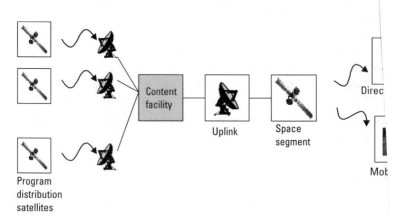

Figure 7.4 Broadcast systems.

international telecommunication satellites. Some of the so-called maritime terminals were also used on land.

The 1990s saw a tremendous change of direction for such systems in terms of global reach, complexity, and bold approaches, as discussed in Chapter 2. Specifically, Iridium, Globalstar, and ICO systems introduced LEO and MEO satellites systems dedicated to this market alone. Such systems broke several new grounds, such as onboard processing, intersatellite links, and partial or complete independence from national networks. Notwithstanding the lack of commercial success of such systems, they managed—albeit at high costs—to advance the system design and technological approaches in many ways.

Figure 7.6 shows a composite diagram for satellite-based mobile systems, currently in use or in different stages of planning and implementation. Significant points to note are:

- While GEO mobile systems are witnessing some degree of resurgence, the types of systems that would replace the currently operating non-GEO systems is still uncertain. Until early 2004, there had still been a possibility that the ICO MEO systems would get completed. However, that does not seem likely any longer. What would follow once the Iridium and Globalstar satellites reach the end of their lives in about 5 to 7 years is also not yet clear.

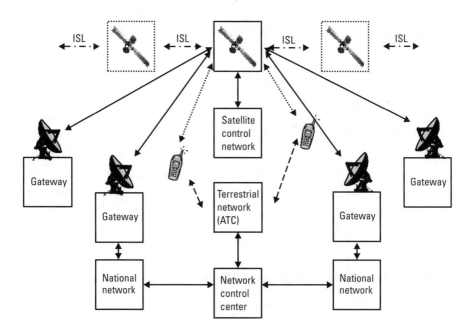

Figure 7.6 Mobile networks.

- The marketing lesson of being fully competitive in all aspects with the terrestrial cellular systems has indeed been driven home. All newer systems are being conceived around handsets either identical to the cell phones or not too different in size.

- In the United States, some of the mobile satellite service (MSS) applicants have been authorized to use parts of their allocated spectrum for operating their own terrestrial networks as well. This decision, known as Auxiliary Terrestrial Component (ATC), is expected to enable the satellite-based systems achieve much better financial returns and customer satisfaction by providing seamless service through a common handset in the entire coverage area.

- Mobile satellites in planning and about to be launched are not focusing just on narrowband voice communication. Instead, they are branching off into mobile videophone and broadband services as well.

Major Engineering Elements

From this brief synopsis of different applications, we identify the following major engineering elements or building blocks:

- Consumer equipment;
- Spacecraft;
- Earth stations;
- Spacecraft control network;
- Content facilities;
- Business and operation systems.

We will now analyze the approaches applicable to each of these elements with a view to meet the objectives of the business strategy. The first three items are covered in this chapter, while the remaining elements will be addressed in the next chapter on operations.

Consumer Equipment Development Program

The list of common elements of satellite-based systems places the consumer equipment right at the top for a good reason. For systems providing direct access to consumers (e.g., broadcast and broadband), the consumer equipment features and the total consumer costs often have a critical bearing on the overall success of the enterprise.

A good and comprehensive survey during the business strategy development phase identifies the service demand profiles against time and sensitivity to the total consumer costs. If the planned service is free on the air (i.e., advertisement supported), the identified consumer cost in the market survey becomes the ceiling that the consumer equipment designer has to meet. If, on the other hand, the consumer cost has a subscription component, then the system-planning phase of the project should have identified an appropriate breakdown between of the total consumer cost between equipment cost and subscription cost. Such a breakdown is often a result of the total project cost optimization, including the space segment, recognizing that a more powerful satellite, for example, can help reduce the cost of the consumer equipment within a certain range. Such optimization will generally fix the front-end sensitivity or G/T of the consumer equipment. We will now examine this aspect in more detail in terms of overall system development as well as the required functional requirements of different engineering elements.

Figure 7.7 shows two different ways of arriving at the functional requirements of the engineering elements. While on first glance the two approaches shown may be expected to give similar answers, in reality they capture two different approaches—not only for the system planning and engineering phases, but often also about the overall organizational approach to business.

Figure 7.7(a) shows what can be called a satellite-centric methodology. Here, the starting point is the spacecraft with emphasis on both the highest capabilities that the spacecraft and the launch industries can provide and fully occupying the available spectrum at the allocated orbital location. Such an approach has been the tradition for communication satellites under the long-standing axiom that the higher the number of transponders in a spacecraft, the lower the cost per transponder. However, for the consumer business, such an approach carries the real danger that when the consumer equipment is developed, the spacecraft development might have proceeded beyond the point of significant changes if they are needed for meeting consumer features and market-driven price targets. As has happened in some recent satellite-centric projects, the management may find itself between the proverbial rock and a hard place halfway through an expensive space segment program. If they decide to modify the spacecraft (assuming that is still feasible), they face high spacecraft modification costs and serious delays in business start-up; if they decide to live with the consumer equipment or costs, they run the different but real risk of serious business shortfall due to the inability to provide what the market requires. Either option can lead to serious shortfalls in returns on investment due to an inadequate match with the marketplace, either in terms of consumer costs or timeframes or both.

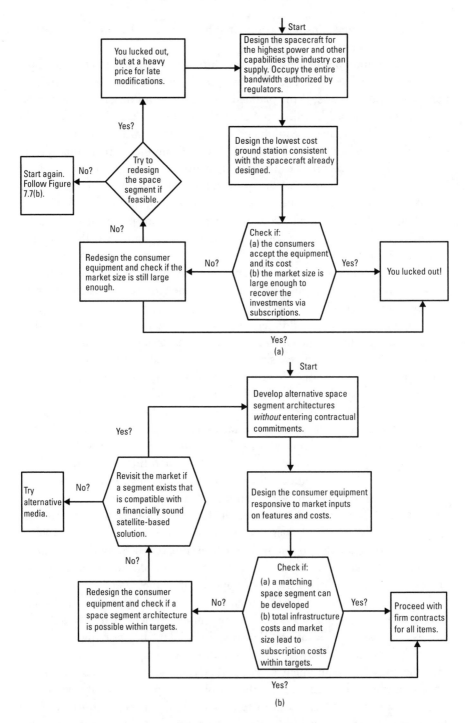

Figure 7.7 (a) Spacecraft-centric approach to consumer business, and (b) consumer-centric approach to consumer business.

In the consumer-centric approach shown in Figure 7.7(b), on the other hand, no contractual commitments are made on the spacecraft until consumer equipment consistent with what the market demand dictates is developed. Only at that stage is the spacecraft configuration firmed up. If, for some reason, such a match is not feasible, the entrepreneur can abandon the use of satellites at a relatively moderate cost. In real life, an indirect implication here is in the choice of program management personnel. The decision makers have to recognize right at the outset that they are in a consumer business and not satellite business per se.

Figure 7.8 shows a more "engineering" look at the appropriate design of overall infrastructure. The pros and cons of the two scenarios in Figure 7.7 can be quantified using this figure. In the satellite-centric approach, you would first fix the key parameters for the spacecraft, whereas in the consumer-based approach, you first determine the G/T for the consumer equipment and then work from the right to the left to design the rest of the key elements, including the spacecraft.

Another parameter that can be the topic of overall system optimization exercise under system planning is the signal compression technique adopted. This is almost equally applicable for both television and radio broadcast systems, although the scale of RF bandwidth are vastly different (of the order of 500 MHz or more for television and only 4 MHz or so per satellite for radio). Assuming the market survey has identified a certain minimum number of channels (i.e., programming choices for the consumer), the choice of compression technology has a

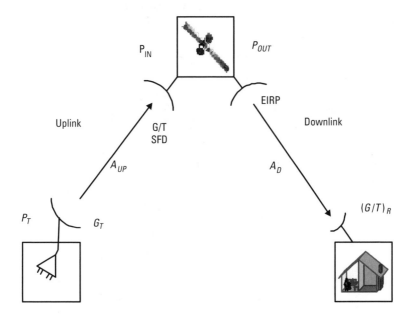

Figure 7.8 Link diagram.

bearing on the satellite capabilities (and hence cost) and to some extent on the digital portion of the consumer equipment. In this regard, it is important to note that the compression technology continues to advance as computing costs drop, and therefore the system as a whole should ideally be capable of taking advantage of these advances during the lifetime of the project. While the space segment can adapt to higher compression efficiency (by providing even more channels or finding alternative use of the spare capacity), it is a tough call for the consumer side because this may lead to a recall of millions of units. In this regard, future systems should consider the adoption of software-derived design approaches wherein the receiver can adapt to changing demodulation and compression techniques by simply downloading an appropriate algorithm [5, 6].

Consumer Equipment Subsystems

We now look at the consumer equipment in a little more detail. Figure 7.9 shows a generic diagram for typical consumer equipment for broadcast systems. The major subsystems are:

- *Antenna module.* This module is directly affected by the spacecraft downlink parameters. For television receivers, the antenna is typically an 18- to 24-inch dish at Ku-band. It has line-of sight visibility with the satellite, and the received signal levels are predictable except for variable attenuation caused by weather. The module has built in one or more low noise amplifiers (LNAs), depending on the number of

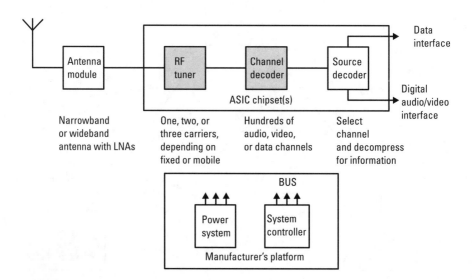

Figure 7.9 Generic receiver architecture.

polarizations and directly connected receivers. The antenna has narrow beamwidth and can typically cover a bandwidth of 500 MHz or higher.

The antenna modules for digital radio systems have functions and sensitivity to spacecraft parameters similar to that for television receivers. However, that is where the similarity ends. The operating frequencies are much lower (L- and S-bands), and the bandwidths are much lower (4 to 20 MHz). The antennas have much broader beamwidths in order to allow reception in a wide range of orientation of the mobile vehicle, not only from the satellite but also from terrestrial repeaters whenever available.

The design challenge is to meet fairly stringent performance parameters at low cost. The digital radio challenges are somewhat higher due to multiple directions of the incoming signals and mobility, often with orientations far from line of sight with the satellite(s) or terrestrial repeaters.

- *RF tuner.* The RF tuner for television is a straightforward RF amplifier and down converter to translate the incoming frequency to a lower frequency where decoding can be carried out efficiently. For digital radio, the tuner can be more complicated. It can be called upon to handle up to three closely spaced RF frequencies from multiple spacecraft and terrestrial repeaters. Because all of these frequencies carry the same information package (with some delay differentials to provide time diversity), only the best frequency or carrier is selected for further processing by the channel decoder.

- *Channel decoder.* The incoming carriers are demodulated and demultiplexed in the channel decoder. While the television receiver has a straightforward quadrature phase shift keying (QPSK) demodulator, for digital radio in addition there is a separate demodulator for terrestrial repeaters often using coded orthogonal frequency division multiplex (COFDM) to counteract multipath-induced signal impairments.

- *Source decoder.* Here, the selected channel or program is decompressed to recover its original configuration. As discussed earlier, this is where continuing advances are still taking place, and a hard-wired solution can become a constraint when competing with systems developed with more recent compression technologies.

- *System controller.* This is where individual manufacturers can provide display and control features tailored to their respective target markets.

- *Power supply.* For portable versions, battery life can be a major factor in the marketability and operating expenses of the receiver.

Key Challenges for Consumer Equipment

Finally, we will highlight the key challenges for the consumer equipment. On the surface, the equipment appears relatively straightforward. Nevertheless, as discussed earlier, it can determine the success of the whole enterprise. Among the key challenges, some apparent and some latent, are:

- The market survey that provides the equipment features and price targets must be as accurate as possible in terms of specific data as well as applicable time frames. Any errors here can permeate the entire project and often determine its very outcome in the marketplace.

- The upfront development effort for the consumer equipment should be a rigorous exercise conducted by a competent group well versed in the related fields. The worst thing one can do is to entrust this effort to spacecraft engineers or organizations. The design and cost disciplines are poles apart in spacecraft teams when compared to consumer equipment development teams.

- ASICs are now an accepted approach in almost the entire range of electronic products. Use of ASICs for consumer equipment is desirable for several reasons: reproducibility of performance, lower production costs, lower current consumption, and higher reliability. However, the induction of ASICs requires a highly disciplined approach in order to realize these benefits in practice. The development process of ASICs is a time-consuming and often quite expensive one. It cannot be compressed too much, nor can any significant changes be made during the process without incurring additional costs and time delays. In most cases, it becomes mandatory to first rigorously develop and test some kind of engineering models with discrete or programmable components before embarking on the ASIC program per se. The underlying technology of ASIC devices is advancing almost continuously and has a direct bearing on the power consumption and to some extent the size of the consumer equipment. However, choice of the very latest technology, generally in terms of the smallest dimension (measured in nanometers nowadays), can lead to higher difficulties in fabrication and yield at least in the initial batches. Recently, the field programmable gate arrays (FPGAs) are also becoming competitive options and in certain applications could provide the much-needed flexibility.

- Equipment packaging and human interfaces can have an important bearing in the consumer acceptance as well as cost. Often, it pays to have professional designers handle this instead of making a "committee decision" inside the project.

- Distribution costs for consumer equipment can often be comparable and sometimes even higher than the ex-factory price for consumer equipment. This must be factored in right from the beginning.

- Unlike long-life spacecraft, consumer equipment components, design approaches, and customer tastes and preferences change on relatively short cycles, typically three to five years at the most. This period should be taken into account in the amortization of development costs as well as in the design of the spacecraft functionalities and flexibility.

Space Segment Development Program

Spacecraft and their associated launches generally represent the largest cost and the critical path in the schedule (*long pole* in industry jargon) of most satellite-based services. An exception could be when the services are planned to be provided by leasing a few transponders from a satellite system already in operation. In consumer-oriented services, it is quite possible that the eventual cumulative cost of content and consumer services could exceed that of the space segment. However, the latter is generally spread over several companies in the consumer electronics field, and the system operator generally shoulders only its initial developmental expenses.

The two main components of the space segment are the spacecraft and its launch services. These two can be procured either separately or as part of an integrated in-orbit delivery contract with one entity, generally the spacecraft manufacturer.

For the spacecraft proper, the program management approaches during the last 50 years have swung from one extreme of total involvement to a totally hands-off one-page financial deal on the other. In the initial stages, the multinational organizations like INTELSAT had to shoulder the burden of advancing the state of the art of the industry through their space segment procurements and therefore had an approach to provide very detailed specifications followed by intensive monitoring of the development and production process of the spacecraft and often that of the launch vehicle itself. This was also to some extent dictated by the obligation to provide visibility to all of the member nations at all stages of the organization's processes. The domestic programs were generally simpler in design as well as management structures and frequently took advantage of the prior developmental investments by the multinational organizations.

The IBSP process described in the following subsections falls roughly between these two extreme approaches of the past. As with all other parts of this process, the principal emphasis is on market responsiveness and not on engineering sophistication for its own sake.

System Planning Inputs

Working closely and interactively with the engineering group, the system-planning function will generally provide the following business-driven requirements as they relate to the space segment portion of the overall infrastructure of the system (see Figures 7.2 through 7.6):

- *Payload capacity.* This may be defined in a variety of ways, depending on the types of services envisioned. The most common unit is the number of transponders of standardized bandwidth, generally 36 MHz or equivalent units. For services operating within relatively narrow RF spectra, such as mobile or digital radio, the total satellite RF spectrum is generally of the order of a few megahertz only. In such cases, the overall payload capacity is generally stated in terms of simultaneous phone calls or audio channels as the case may be.

- *Coverage requirements.* These are directly dictated by the business area and indirectly by the regulatory requirements to limit the radiated power within certain limits outside the coverage area. For broadcast systems, receive coverage requirements are less stringent than the transmit ones.

- *Compatibility with the ground segment.* This can vary over quite a wide range, as discussed earlier for different types of infrastructures. In general, with medium to large telecommunication antennas on the ground, the emphasis in the spacecraft is on performance and capacity, whereas for broadcast systems, as we have seen, the major emphasis is on providing the highest power from the satellites. Mobile systems in many ways straddle both of these categories.

- *Spacecraft performance requirements.* A strong and interactive planning activity will provide performance requirements driven by the overall end-to-end market-driven requirements rather than technology-driven ones.

- *Number of spacecraft and deployment schedules.* These are directly mandated by the market needs, and often the schedule can be a crucial element of success, as discussed in earlier chapters, including Chapter 2.

Table 7.1 tries to capture how the common key parameters can be specified for typical satellite-based infrastructures. Table 7.2 is a matching table to show the corresponding impact on principal spacecraft parameters. The next section discusses such aspects in a little more detail.

Generic Spacecraft Architectures

Seen from the market and application perspectives, spacecraft for different applications appear quite different from one another. Even physically, significant differences can be seen. Thus, a telecommunication satellite for international applications may have a number of antennas of different sizes and shapes, some with very complicated feed arrays. On the other hand, a domestic satellite for program distribution may have just one transmit antenna. And, finally, a mobile services satellite may have a couple of extremely large mesh antennas and possibly external thermal radiators and so on. Nevertheless, from the perspective of business and programmatic management, it is possible to identify some common approaches facilitated through some generic architecture charts.

Payload

Figure 7.10 shows a generic architecture of a typical satellite payload in terms of common major subsystems. For each of the building blocks, its function, range of complexity, and impact on the program are captured directly below it. Let us briefly discuss each of the modules in turn.

For broadcasting systems, the receive antennas and receivers are relatively straightforward because they often have just one link to the normal uplink facility and perhaps another to a backup site. An exception can be in situations

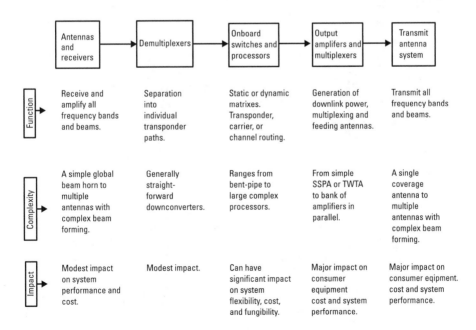

Figure 7.10 Generic spacecraft payload architecture.

Table 7.1
System Planning Inputs to Engineering for Spacecraft

	Telecom Systems (Figure 7.2)	Broadcast Distribution Systems (Figure 7.3)	Broadcast Systems (Figure 7.4)	Broadband Systems (Figure 7.5)	Mobile Systems (Figure 7.6)
Capacity	Number of 36-MHz transponders or equivalent	Number of 36-MHz transponders or equivalent	Number of transponders or total digital throughput	Downlink digital throughput and number of simultaneous uplinks	Total digital throughput, number of channel slots, or simultaneous conversations
Coverage	Up-down link business areas	Up-down link business areas	Downlink business area plus uplinks stations	Up-down link business areas	Up-down link business areas
Transmit EIRP	EIRP for each transponder	EIRP for each transponder	EIRP for each transponder or total EIRP	EIRP per transponder	EIRP per cell and total capacity
Receive G/T	Specified to match existing or planned earth stations	Specified to match existing or planned earth stations	To match content/ uplink station parameters	To match content/uplink station parameters	Match maximum power from the handset
Connectivity	Static, switched TDMA or traffic switching	Generally static connectivity matrix	Generally bent pipe	Bent pipe or onboard processing	Onboard or ground processing
Regulatory Contraints	Sharing constraints on both up-down links	Sharing constraints on both up-down links	Minimal if in exclusive bands	Sharing constraints, earth station size	Constraints on interference to adjacent operators
Performance	Total noise contributions	Limits on quality impairments	Quality limits, fade margins	Bit-error rate performance	Quality, availability limits

Table 7.2
Impact of System Planning Inputs on Key Spacecraft Parameters

	Telecom Systems (Figure 7.2)	Broadcast Distribution Systems (Figure 7.3)	Broadcast Systems (Figure 7.4)	Broadband Systems (Figure 7.5)	Mobile Systems (Figure 7.6)
Total Power	Number and power of transponders, onboard processing complexity	Number and power of transponders	Number and power of transponders	Number and power of transponders, onboard processing complexity	Power per circuit and onboard processing complexity
Total Mass	Total power, size of antennas, their feed complexity, and propulsion scheme	Total power, size of antennas, propulsion scheme	Total power, size of antennas, propulsion scheme	Total power, size of antennas, onboard processing, propulsion scheme	Total power, size of antennas, onboard processing propulsion scheme
Complexity	Number of transponders, onboard processing, antenna systems	Number of transponders, antenna systems	Number of transponders, thermal design, antenna systems	Number of transponders, onboard processing, thermal design, antenna systems	Onboard processing, antenna size, power, and thermal design
Cost	Power, mass, complexity, development and qualification status	Power, mass, development and qualification status	Power, mass, development and qualification status	Power, mass, development and qualification status	Power, mass, complexity, development and qualification status
Schedule	Subcontracting, development and qualification needed, testing facilities	Subcontracting, development and qualification needed, testing facilities	Subcontracting, development and qualification needed, testing facilities	Subcontracting, development and qualification needed, testing facilities	Subcontracting, development and qualification needed, testing facilities

where the uplinks can originate from a number of different places, either sequentially or simultaneously. (An example of the latter would be the World-Space radio broadcast satellite discussed later in Chapter 10.) Even in those situations, generally the receive antenna subsystems are simple and represent a small fraction of the total costs. For telecommunication satellites, however, especially for international applications, the receive antenna subsystem is often a mirror image of the transmit antenna subsystem. It can be equally complex in terms of multibeam capabilities and can be an important element in the overall costs and program schedules.

The demultiplexers are generally simple, as their function is quite straight-forward translation of the incoming frequency to a lower frequency band for further processing. However, in cases where digital techniques are used (e.g., mobile applications), there can be higher complexity, and this function can be integrated with the processing function itself.

Onboard switching and processing can cover the full range of complexity, starting from simple static switches to all the way to extremely complex processors. The *bent-pipe*, or transparent, satellites, still the work horse of the industry, generally have simple static switches operated manually from ground command to provide the desired connectivity between different uplink and downlink pathways. Such switches are independent of the type of applications or modulation/access techniques adopted—hence the name *transparent*. At the other extreme, you find very complex processors resembling modern telephone exchanges on the ground. They are much lighter but far more expensive! Such processors enable much higher overall capacity levels to be achieved in applications such as geomobile systems within relatively narrow RF bandwidths. Such processors can also include complex antenna functions for creating hundreds of beams, both in the uplinks and downlinks. In spite of adoption of the very latest semiconductor technologies in the ASICs, such processors can require thousands of watts of onboard direct current power. And, finally, such processors can be very application specific and therefore from a business perspective these can be high-risk investments if the related demand has too much volatility.

The output amplifiers can be quite complex and critical from several perspectives. They obviously account for the bulk of the total satellite power needs and hence control the overall cost. Because failure in such amplifiers, whether traveling-wave tube amplifiers (TWTAs) or solid-state power amplifiers (SSPAs), can often mean total link failure, appropriate redundancy schemes are necessary, often leading to higher complexity, mass, and cost. Telecommunication satellites generally have a large number of transponders, each with medium-power amplifiers with their own or shared redundancy schemes. In broadcast systems, the spacecraft have fewer transponders but each has higher power. In the extreme case of satellite radio, the spacecraft may have just one transponder made up of a large number of paralleled amplifiers—a virtual microwave power station, if you will!

The transmit antennas obviously have to cover the business areas while at the same time meet strict requirements for not exceeding specified levels outside such areas. These antennas are therefore more complex than the receive antennas, particularly for broadcast systems. For telecommunication systems, as discussed earlier, receive and transmit antenna systems can have similar complexity with some differences. The transmit antennas have to handle much higher power levels and are generally physically larger due to lower operating frequencies. For mobile satellite systems, transmit (and receive) antennas can be extremely large, requiring very elaborate packaging and deploying challenges.

Bus (or Platform)

Figure 7.11 is a similar chart for the bus or platform for modern spacecraft. As can be expected, the spacecraft bus is much less application specific, and therefore a much higher degree of standardization can be expected for the different subsystems. Along the lines of the previous diagram, Figure 7.11 identifies the functions, complexity, and impact of each of the major subsystems.

The power subsystem is obviously an extremely important item and is generally the principal differentiator between different buses and manufacturers, as

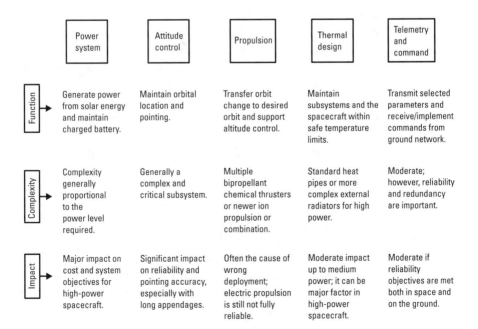

	Power system	Attitude control	Propulsion	Thermal design	Telemetry and command
Function	Generate power from solar energy and maintain charged battery.	Maintain orbital location and pointing.	Transfer orbit change to desired orbit and support altitude control.	Maintain subsystems and the spacecraft within safe temperature limits.	Transmit selected parameters and receive/implement commands from ground network.
Complexity	Complexity generally proportional to the power level required.	Generally a complex and critical subsystem.	Multiple bipropellant chemical thrusters or newer ion propulsion or combination.	Standard heat pipes or more complex external radiators for high power.	Moderate; however, reliability and redundancy are important.
Impact	Major impact on cost and system objectives for high-power spacecraft.	Significant impact on reliability and pointing accuracy, especially with long appendages.	Often the cause of wrong deployment; electric propulsion is still not fully reliable.	Moderate impact up to medium power; it can be major factor in high-power spacecraft.	Moderate if reliability objectives are met both in space and on the ground.

Figure 7.11 Generic spacecraft bus architecture.

seen from the application or business perspective from the outside. Over 50 years of evolution, the power-generation capabilities of spacecraft have grown from a few tens of watts to up to 20 kw today. This has been achieved by progressively higher efficiency solar cells and the availability of higher mass capability of modern launchers. While great care is taken to provide adequate redundancies and fail-safe practices, random failures in power subsystems still remains one of the principal reasons of in-orbit failures.

Attitude control of spacecraft is a very critical background function for modern spacecraft. It is responsible for stabilizing the spacecraft and for maintaining accurate pointing of antennas toward specified points on the ground under all conditions of variations in gravitational forces and solar pressure on the spacecraft as well as movement of large flexural appendages such as antennas and solar arrays. This is achieved through small thrusters forming part of the propulsion subsystem. As spacecraft have grown in mass, size, and number of appendages, the attitude control functions have also become more complex.

The propulsion subsystem has two major and separate functions. The first one is to assist in the transition the spacecraft from an elliptical transfer orbit just after release from the launch vehicle to the desired orbit and operating location. The second relates to attitude control. Most satellites achieve both of these functions through a common system where thrusters of appropriate sizes achieve the required thrusts for specified durations by mixing two chemical compounds. The transfer orbit operations are carried out by a much larger thruster, frequently called *apogee motor*, a carry-forward of a name from solid motor days. The attitude control functions are performed through several redundant thrusters placed outside the main body of the spacecraft in order to provide the ability to control movements in all planes. In a well-designed spacecraft with adequate redundancies, both the launch mass as well as the useful life of the spacecraft can be a direct function of the total quantities of the bipropellant fuels carried onboard. Among the techniques being used to at least partially remove this life limitation are a variety of electric propulsion devices under development for several decades. Electric (or ion) propulsion devices can provide the same total thrust for a much less physical mass of the propellant, generally xenon.

A typical spacecraft encounters a very wide range of body temperatures. This is particularly true for the three-axis spacecraft, a configuration now responsible for all but a small fraction of total spacecraft. In the absence of any atmosphere surrounding the spacecraft, different parts of the spacecraft body can have relative temperatures differing by several tens of degrees or more. The exact differential at any given time is determined by the orientation toward the Sun and power dissipations inside the spacecraft. The thermal control system maintains each subsystem temperature within its qualification range at all conditions of load and Sun orientation. This is achieved by monitoring the total thermal environment through a large number of thermal sensors throughout the spacecraft—heat-pipe systems that reduce temperature differentials between different spacecraft panels and switchable heaters at several key locations. Very-high-power spacecraft may also use in addition external radiator systems.

Most modern spacecraft do not operate autonomously beyond a certain period, typically in terms of hours. All of the key parameters are monitored continuously from the ground through the telemetry, command, and ranging subsystem. These systems are highly redundant and provide for sending down the sensor, switches, and other control information and receive commands for execution for a range of functions under the control of the ground network. More details will be discussed in the chapter on operations.

Spacecraft Procurement Process

As mentioned earlier, the procurement processes adopted by various organizations vary widely. Government processes, including those for the military, are generally quite elaborate in part due to the very specialized and complex functional needs of such spacecraft. Apart from confidentiality requirements, such programs quite often push the state of the art to its limit in order to meet certain strategic and tactical objectives. The procedures adopted reflect these necessities with considerable variations in the procurement and contracting practices adopted in different parts of the world.

For commercial systems as well, there is a very wide range of practices, as noted earlier. Some of these differences can arise from national industrial policy requirements. The nature of the spacecraft can also dictate to what extent some of the intermediate steps might be either shortened or strengthened. All other factors being equal, a program requiring considerable new development work often by a group of contractors can require more elaborate procedures. On the other hand, replacement spacecraft or spacecraft similar to those already in orbit may adopt highly simplified procedures, particularly if the customer and the manufacturers have longstanding relations and understand each other's practices and perspectives.

Once again, in order to put our arms around such a wide range of procedural and administrative procedures, we will identify certain common steps to capture their respective nuances and relative importance from business and programmatic perspectives.

Figure 7.12 shows the typical procurement steps for a commercial spacecraft program. Once the system planning teams in the IBSP have developed the general interface requirements, the mechanism for getting started is often a requirements statement (or something equivalent). Such a statement is not too long; neither is it an abbreviated version of the final spacecraft technical specification. Rather, it is a short summary of the system-planning requirements that can serve many purposes. As a starter, it can be used to reconfirm within the

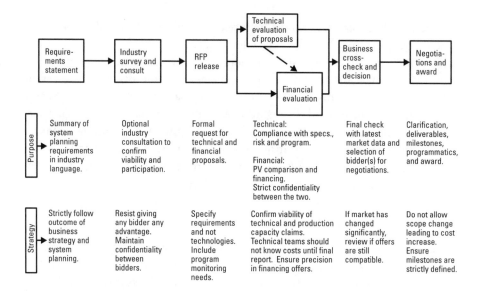

Figure 7.12 Spacecraft procurement process.

company that all parts of the organization—and certainly those within the IBSP—are still satisfied with the general direction upon which the engineering is about to embark. It can also be useful to ensure support of the company board, governing council, and other stakeholders in the venture as applicable. Quite often, such coordination not only builds organizational program support but also leads to certain improvements and enhancements, particularly with regard to changing external environmental factors such as market-related and competitive landscapes.

Once this team coordination is completed, such statements can be shared with the industry in order to illicit interest and receive general feedback on the compatibility of what is needed with what industry is capable of delivering without requiring too much new development. (Such an industrial process can be carried out for all other procurements as well and need not be limited to spacecraft procurement.)

Armed with all of this internal as well as external coordination, the engineering and procurement teams embark on the preparation of the formal request for proposals (RFPs). Unless unique circumstances dictate otherwise, open global competition almost always provides the best outcome in terms of the optimum combination of performance, price, and schedule.

A fair and thorough evaluation of proposals does require discipline in terms of following related procedures. Irrespective of the size of the procurement, it always pays to perform the technical evaluation without the concerned experts being aware of the prices quoted. The financial evaluation is generally shorter in duration and can therefore start a little later, certainly after any clarifications in the technical offers have been obtained if needed. In the financial evaluation, very often the engineers can be called upon to assist in properly understanding complex financial offers. However, this process should be very carefully carried out without breaking the "firewall" between the two evaluations.

At the end of the evaluations, it is not uncommon to find all bidders technically compliant, with mostly qualitative differences between them. Significant differences in launch mass can indeed be a differentiator; however, this generally translates to a financial factor through launch cost differences. Therefore, the choice after completion of evaluation quite often comes down to price and perhaps schedule differences only.

Before the final contract is awarded, two more important steps are required. The first is a final crosscheck with marketing needs and compatibility with the whole IBSP process. The second is detailed negotiations with the top bidder for the deliverables, milestones, and program oversight requirements. Until this stage, it is a wise approach to maintain at least two bidders still in the race in order to provide a backup and prevent the top bidder from exploiting his status as the front runner.

Oversight of Spacecraft Development Program

Monitoring of spacecraft programs through on-site personnel is as old as space technology itself. In the beginning, almost all programs were really developmental programs in practically every respect, ranging from bus and payload components and subsystems and even manufacturing and testing processes. As a result, the contractual specifications were often starting points rather than the assured final outcome of the complicated process of putting the spacecraft together over a number of years. Program slippages, mid-course contractual changes, and price increases were almost the order of the day rather than an exception. Accordingly, major customers recognized a genuine need for a close day-to-day interaction with the designers at the manufacturers, rather than being presented with a fait accompli situation after significant mid-course corrections had already been made. This led to the practice of on-site monitoring teams that continues even today in some form or other.

On-site monitoring of programs of adequate complexity can in principle be beneficial to both sides. For the customer, its engineers are in a position to monitor progress as well as setbacks as they happen rather than through an after-the-fact-cleaned-up report. Specifically, they are in a position to monitor details of qualification and testing programs at all levels of the program. For the manufacturer, its personnel are able to get continuous feedback from the customer on new approaches rather than be forced to wait until formal concurrences can be obtained via contractual correspondences.

Obviously, there are pitfalls and disadvantages as well. Apart from additional personnel costs, the customer runs the real risk of its staff, after a prolonged stay at the manufacturer's premises, losing their critical *customer perspective* and unknowingly beginning to think more and more like the manufacturer. Through the inevitable give and take between the monitoring team of the customer and the program team of the manufacturer, there can be compromises made that may not be in the larger interest of the customer management. These potential shortfalls can often get accentuated if the monitoring is contracted out. While professional contract monitors are generally very experienced professionals, they may not be at all tuned to the strategic and business policies and practices of a customer located far away.

Seen from the manufacturer's perspective, there are certain disadvantages, too. Its program management team may begin to feel a bit cramped by the perennial presence of customer staff at all meetings, especially those of strictly internal nature. The mid-level managers can begin to lose their initiative, as they often tend to cross check almost everything with the monitoring team manager first before even bringing their own decision chain in the loop. Above all, a competent program team can begin to feel micromanaged by the customer.

Overall, program oversight can be beneficial if it is kept at a moderate level. Some of the general guidelines that can make such processes successful at reasonable cost include the following:

- Program oversight by an on-site team is no substitute for choosing a competent contractor. The customer should resist micromanaging the contractor and accepting some of its responsibilities in the name of improving quality and reducing risks.

- The monitoring personnel must rotate through the parent organization at periodic intervals rather than be away too long at industrial sites, or they will no longer accurately reflect the customer's strategic objectives behind the program.

- The customer should resist suddenly increasing oversight if the program runs into difficulties. The resident teams should maintain their stature under all circumstances.

- A competent manufacturer can feel cramped and micromanaged by an intrusive program team. The time to prevent such a situation is at contract award time and not when things are going wrong during the program.

- Government regulations often constrain the manufacturer from revealing certain details of their processes and technologies. Such constraints should be clearly stated at the outset in order to avoid unpleasant situations downstream.

- The manufacturer, like any other business entity, has a right to have totally internal meetings and documents. The monitoring team must respect such normal privileges.

- Finally, the greatest danger is to have an idle specialist in the monitoring team. He or she will look for problems even when there are none. The bulk of the resident staff should be programmatic experts who should bring in specialists only when warranted.

Launch Services

Unlike the spacecraft development program, the launch services are generally a straightforward procurement action. Except in rare situations, no specific development activities are needed to match a particular spacecraft's needs. Of course, the spacecraft should be designed so that it is compatible with at least one available launch vehicle.

Launch services can be either procured either directly or through the spacecraft manufacturer. In the latter case, the deliver of the spacecraft is in orbit and all responsibilities associated with a timely launch, including launch insurance, belong to the spacecraft manufacturer. Such an option can be attractive for ventures that do not need to launch multiple spacecraft at short intervals, as it can lead to savings in personnel costs. However, an in-orbit delivery contact can have some pitfalls that should be recognized, such as:

- The contract should give adequate visibility to the customer about launch backups in case the primary launch vehicle offered has encountered a recent failure. Late delivery penalties are generally small and often have enough loopholes to be almost ineffective.

- The spacecraft should have clear acceptance criteria, both ex-factory as well as in orbit.

- The customer should use its own ground network and staff for final acceptance in orbit.

Earth Stations

We have already reviewed in some detail the consumer equipment, which is really an earth station, albeit a very small one. There are other types of earth stations, of course, as we can see in Figures 7.2 through 7.6 on different types of infrastructures. The different types can best be categorized by the types of applications and infrastructures. Using the same broad classifications as before we will can now highlight the major types involved.

Telecom Networks [Figure 7.2(a)]

Medium size gateway antennas and VSAT systems are the two categories here. The former emphasize capacity and reliability, whereas for VSATs, cost is a main driver because they are used in large quantities.

Connectivity Satellites [Figure 7.2(b)]

These are generally part of international networks across oceans providing connectivity to a large number of earth stations. Such stations are generally the gateways of individual countries, and their numbers may vary from one to under 10 per country, depending on the number of satellites accessed. Such antennas carry a very large amount of traffic and therefore emphasize efficiency and reliability even more than the gateways discussed earlier. Some of these antennas are in operation for several decades and are much larger is size than would be required for the more powerful satellites in use today for such applications.

Program Distribution Systems (Figure 7.3)

These are one-way links in contrast to the two-way telecommunication links discussed earlier and are therefore simpler. The transmit uplink antennas are generally colocated with content or programming network facilities. Once again, reliability is the main driver here, and redundant installations are generally the norm.

Direct Broadcast Systems (Figure 7.4)

The uplinks here are generally similar to those for program distribution. We have already discussed the consumer antennas.

Broadband Systems (Figure 7.5)

Once again the uplink stations are similar to the types discussed earlier, except that their content is now principal Internet connectivity on demand. The user systems can be either at home, as shown in Figure 7.5, or at businesses. For home applications, the receive systems are similar to the consumer equipment discussed earlier; however, the transmit equipment is an important addition. Within strict cost constraints, these small transmitters have to meet typical uplink constraints of small VSATs. For businesses, larger facilities akin to VSATs are used. These provide higher throughputs in both directions, and the design approaches are more like those adopted in professional systems.

Mobile Networks (Figure 7.6)

The country gateways for mobile networks are similar to those for telecommunication systems. However, the user equipment has seen significant advances over time. In the initial phases, the mobile terminals were essentially VSATs mounted on ships with necessary attitude stabilizing systems to ensure steady pointing toward the satellites. As the satellites have become more powerful, smaller portable terminals have been introduced for use on land, sea, and even aircraft. The well-known LEO/MEO systems ushered in the era of handheld consumer equipment, no longer labeled as earth stations or VSATs. This trend is likely to continue with the objective of making the satellite phones as compact and convenient as cellular phones.

Development and Procurement Issues

By and large, earth stations by themselves are becoming highly standardized, and little or no development is required by the customer. The principal management focus is therefore on timely acquisition of reliable systems. However, there are areas where development and programmatic efforts of the types discussed earlier for consumer equipment and spacecraft are still necessary. One such example is the broadband consumer equipment still going through considerable development and evolution. Compression techniques are also continually advancing, and therefore careful choice and often dedicated development work could be necessary.

References

[1]   http://www.wildblue.com.

[2]   Thesling, W., et al., "Two-Way Internet over iPSTAR Using Advanced Error Correcting and Dynamic Links," *20th AIAA International Communications Satellite Systems Conference*, paper 2002-1944, May 2002.

[3]   Swakpan, T., "The IPSTAR Broadband Satellite Project," *21st AIAA International Communications Satellite Systems Conference*, paper 2003-2206, April 2003.

[4]   Sarraf, J., "The Spaceway System: A Service Provider's Perspective," *IEE Seminar on Broadband Satellite: The Critical Success Factors—Technology, Services, and Markets*, October 17, 2000, pp. 15/1–15/6.

[5]   Bing, B., and N. Jayant, "A Cellphone for All Standards," *Spectrum IEEE*, May 2002, pp. 34–39.

[6]   Forbes, J., and M. Wormley, "A Handheld Software Radio Based on the iPAQ Hardware," *2003 Software-Defined Radio Technical Conference*, Orlando, FL, November 2003.

8

System Operations

System operations include all activities, tools, and facilities necessary to run an enterprise once all of the infrastructure elements have been completed and the services to the customers are about to start. The actual mix and complexity depends to a certain extent on the kind of services being provided. As a minimum, this group is responsible for the infrastructure and for carrying out the day-to-day business operations. Depending on the nature of the services, this group can also be responsible for the aggregation and creation of the content being distributed through the system. Above all, this group is the long-term interface with the users and the provision of an efficient customer service. As part of this activity, system operations provides valuable feedback on the QoS perceived by the customers as well as suggestions for new features and future evolution of the system. Figure 8.1 summarizes the major responsibilities and interdependencies for system operations as part of the IBSP.

Infrastructure Responsibilities

The two common infrastructure elements are the satellite control network (SCN) and earth stations. In addition, there can be terrestrial networks for certain services.

SCN

This is a critical activity responsible for the control and safety of the space segment. It can be an in-house function, or it could be contracted out to companies with the necessary expertise, experience, and facilities. In the latter case, it is

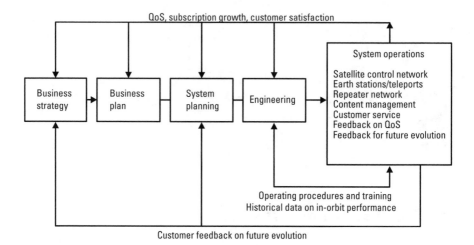

Figure 8.1 Key responsibilities of system operations.

important to establish clear and quick communication channels between the contractor and the operator with unambiguous decision procedures for handling any serious emergencies in the space segment.

As a minimum, the SCN for a single satellite in orbit will consist of two telemetry, telecommand, and ranging (TC&R) earth stations and a control center. The two TC&R stations must have good uninterrupted visibility at all times for the satellite being monitored. While a single such station can provide all the functionalities, the second station provides redundancy against unforeseen shutdowns and improves the accuracy of ranging and velocity measurements for the satellite. Each station has redundant data links to the control center. For simple networks, these earth stations can be unattended. The control center is generally part of system operations unless the whole SCN function is contracted out. It has its own backup facilities. National regulations often require the control center to be geographically located in the country responsible for the registration of the corresponding orbital location with the ITU.

It is quite feasible and cost effective to monitor multiple satellites from a common control center either for a single operator or for multiple operators. In such networks, there are either two dedicated TC&R antennas per satellite or one dedicated antenna per satellite and one larger, fully steerable antenna with the ability to access any of the satellites being managed. This trend has led to large networks having multiple antenna sites around the world.

The principal functions of a typical SCN include the following:

- *Transfer orbit operations following launch.* Many multisatellite control centers are equipped with adequate facilities and expertise to support the monitoring and commanding of a spacecraft as it leaves the launch vehicle and transitions from the transfer orbit to its final orbit and operating location. These operations are generally quite intensive and require participation by a bank of experts, including those from the satellite manufacturer, covering all disciplines and subsystems in the spacecraft. Barring a serious hardware malfunction during the launch phase, such operations are carried out against a carefully scripted transition plan aimed at maximizing the remaining fuel onboard the spacecraft and hence its lifetime for the operational phase.

- *In-orbit testing (IOT)*. This is carried out before the start of operations and involves performance and functional *check out* of the spacecraft platform and the payload. Depending on the coverage configurations, it may be necessary to tilt the spacecraft in several directions to confirm antenna performance and to cross check all of the major pathways through the satellite payload via the onboard switches and processing. The SCN control center plays a critical central role in such tests, ensuring that at no stage the spacecraft is over stressed or placed in any risky situation. With increasing use of in-orbit delivery contracts, these IOTs are also the basis of acceptance of the spacecraft by the customer and the establishment of a reference point for any in-orbit incentives going forward.

- *In-orbit operations*. These operations involve a 24 × 7 surveillance of the health of the spacecraft and certain periodic routine functions. Modern control centers are equipped with specialized software that enables efficient surveillance of multiple spacecraft through intelligent monitors and preset alarm limits for key parameters of each spacecraft. The key routine function is maintenance of the spacecraft within the contractual and regulatory position limits in the east-west and north- south directions for geostationary satellites. Barring any anomalies in the related propulsion subsystems, such maneuvers are generally preprogrammed to be carried out at specific intervals for each spacecraft.

- *Other responsibilities*. Depending on the size and configuration of the system, the SCN is also called upon to carry out several other important functions as needed. These include live transition of traffic from an older spacecraft to a newer one at the same location, movement of spacecraft from one location to another, and reconfiguration of payloads through antenna repointing and transponder switching. The SCN should also have appropriate contingency plans for the spacecraft as well as for the SCN itself. Some large global SCNs also lease their networks for supporting launches of other networks.

In order to efficiently fulfill these responsibilities, a good SCN must have several background capabilities. The most critical one is access to the right experts as needed. This is generally achieved by on-call rosters with remote computer terminals duplicating the operator displays.

An equally important capability is up-to-date spacecraft documentation and procedures. It is traditional to talk about *gray beards* in the aerospace industry to underscore the importance of good institutional memory and record of past experiences, both good and bad. While in some situations this phrase may be overkill, it is indeed important when it comes to minimizing the risks in an SCN. With spacecraft lifetimes extending to 15 years and beyond and with periodic reorganizations in the industry, it is quite possible that the industry engineers who originally designed and produced the spacecraft are no longer accessible in times of technical emergencies during the lifetimes of the relevant spacecraft series. It is therefore mandatory that the engineering groups, while handing over the spacecraft to system operations, provide a complete and exhaustive library of all relevant documents.

The SCN personnel and the on-call engineers should have access to past and current historical data. This has two components. The first is subsystem performance data starting from its qualification, production, and IOT, and updated through in-orbit experience. The second component relates to the data on the dynamic behavior over time of the spacecraft bus and the payload. All of the historical data is generally stored close to the SCN and should be electronically accessible by the experts in the SCN without any bureaucratic procedures or delays.

Training of SCN personnel in the functioning, history, and operating procedures of different spacecraft under monitoring is an important activity. In addition to good documentation, such training is generally aided by spacecraft simulators. As the name implies, these simulators enable the staff to see on their desktops the dynamic behavior of the platform as well as the payload. The payload simulators are also useful in confirming the end-to-end performance of current and new services.

Figure 8.2 shows a generic diagram for a typical multisatellite SCN. Two antenna clusters are shown, each with one fully steerable TC&R antenna. The data lines from the TC&R stations to the control center are fully redundant in order to ensure the continuity of operations. If such networks have a sufficient number of antennas around the world, they can also manage launch missions.

Before we leave the topic of the SCN, it is important to underscore the inherent risks involved. With the advances in attitude control and other related technologies onboard, most modern spacecraft have built-in fail-safe modes from which it is feasible in most situations for knowledgeable experts to bring the spacecraft back to its nominal operation and orientation in space, often without any long-term impact except perhaps some extra fuel consumption. Accordingly,

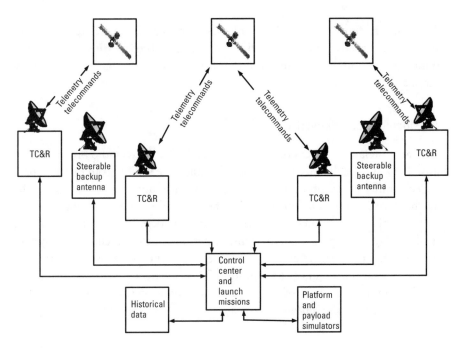

Figure 8.2 Generic SCN.

the number of instances that there was a total loss of spacecraft has declined substantially. Nevertheless, despite all of the technological advances and sophistication, the importance of having access to experienced engineers with the right training and demeanor to handle emergencies in a calm and professional manner cannot be overstated. This factor can and should continue to play an important role in a decision to have in-house or contracted-out SCN facilities.

Earth Stations

As we saw in Chapter 7, there is a wide range of earth stations, depending on the applications. Some of these stations can be the responsibility of system operations.

For fixed telecommunications, the earth stations have traditionally been part of the user organizations, generally the national telecommunication providers. However, this is changing to a certain extent with the increasing popularity of multiantenna facilities, often referred to as *teleports*. Teleports provide access to multiple satellites and other media as well. Independently owned teleports may also provide competitive access to different satellite systems.

For mobile satellite systems, the earth stations that are part of system operations are the gateways providing connectivity to national and regional telecommunication networks. Some of these gateways also act as teleports and can access alternative media as well. In the case of broadcasting systems, the uplink stations and the stations for monitoring system performance are generally part of system operations.

Terrestrial Repeater Networks

These are of relatively recent origin and are used with digital radio broadcast systems. They are installed in urban areas with high-rise buildings in order to provide continuity of reception to mobile users. Typical numbers in an area can be anywhere from 10 to 50, depending on the size and nature of the city and the elevation angles to the satellites. These repeaters work either with the broadcast signal itself or through a separate transponder on another satellite. The system operations has to ensure at all times the integrity of such feeds as well as the satisfactory operations of all of the repeaters. This is generally achieved through a combination of real-time telemetry and automatic periodic polling of key parameters of all of the repeaters.

Content Facilities

The majority of satellites around the globe are still in the business of leasing or selling transponder capacity, either for television and radio distribution or for a range of telecommunication services. However, with the growth of direct consumer services, satellite companies are also getting into the content business through different degrees of vertical integration with the traditional content providers. Currently, there are two main groups of such services: television and radio. In addition, a smaller but growing market is related to providing content as part of Internet services or specialized data.

Direct Television Broadcast

In many parts of the world, direct television broadcast systems based on satellites are selling subscription services based on specific ensembles of programming or content. In most cases, this content is aggregated by the system operations, largely through other independently owned television distribution satellites operated by other organizations. The television broadcast satellite companies do not themselves create any custom programming, but this may change in the future (see Figure 7.4). However, they could have exclusive contracts with certain groups of programming (e.g., sports).

The responsibilities of system operations for such content-related activities include:

- Commercial arrangements for specific content feeds;
- Encoding and multiplexing of the selected programs in a composite signal suitable for broadcast by the uplink station;
- Management of the subscription systems through proprietary conditional access systems.

Direct Radio Broadcast

These are relatively recent systems, covered in more detail as case studies in Chapter 10. Unlike the television broadcast systems, most of these systems rely to a considerable extent on proprietary content put together in house. While this may also change in the future, this trend appears to have been driven by the need to distinguish a new subscription service from the traditional AM/FM free-on-the-air broadcasts.

In-house content generation is largely based on commercially available digitally recorded music, with some live broadcasts by visiting artists. What is unique about these radio channels is the playlist put together by experienced specialists. In addition, the availability of several tens of channels on one broadcast also allows the content managers to provide comprehensive ensembles or *neighborhoods* to meet the tastes of most listeners.

In terms of facilities for such content, the two U.S. systems operate perhaps the most advanced set of studios in their premises for content development. Figure 8.3 shows an example of such a studio [1]. The WorldSpace system creates content in different locations and combines them remotely as needed for different uplink stations. Content aggregation and generation is certainly a major activity for radio broadcast systems. In addition, the other functions mentioned earlier for television broadcasts are also necessary.

Broadband Services

Such services are still evolving, and the scope of the associated operational facilities varies quite a bit among the different players active in this field. Figure 8.4 from [2] shows different services being envisioned and offered in various combinations. As an example, simple two-way access broadband access could be offered to SOHO and small/medium enterprises (SME) types of users. This could also be augmented with caching capabilities at the hub and at corporate headquarters. Multicast capabilities with live data streaming could also be added as needed.

Reference has already been made in earlier chapters to the Spaceway, an all-Ka-band on-demand packet-based connectivity services broadband system, now expected to be operational in 2005 [3]. The operational control of the IP

Figure 8.3 Typical digital radio studio. (*From:* [1]. © 2004 American Institute of Aeronautics and Astronautics, Inc. Reprinted with permission.)

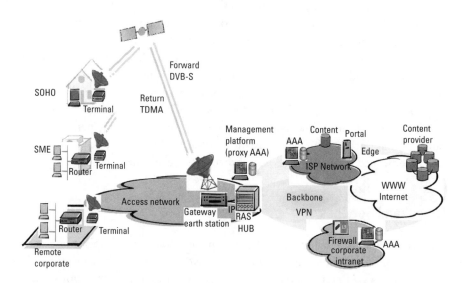

Figure 8.4 Two-way broadband access. (*From:* [2]. © 2003 Alcatel Space. Reprinted with permission.)

network, including IP-address management, is planned to be carried out via a network operations and control center (NOCC). Figure 8.5 from [3] shows a top-level interface diagram for such a function. In general, such network operations centers are called upon to handle tasks that include network management, bandwidth management, gateway, and user-terminal management.

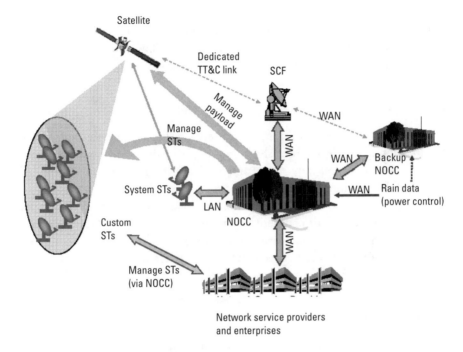

Figure 8.5 SPACEWAY operations facilities. (*From:* [3]. © 1999 Hughes Network Systems, Inc. Reprinted with permission.)

Apart from their two-way modes, the broadband services also differ from broadcast services in terms of their much stricter QoS requirements. This is particularly true with the increasing provision of voice over IP (VoIP) telephone services over broadband links. Such requirements can often become the controlling factor in the design, manning, and operation of the related operational facilities. Reference [4] provides a recent extensive survey of QoS as applicable specifically to satellite broadband networks.

Business Operations

Three common and important components of business operations are as follows.

Billing and Subscription Management

The details of this activity will vary substantially with the kind of services provided. Systems in the business of leasing capacity have the bulk of their business through medium and long-term bookings, with some capacity for short-term or occasional use. For direct broadcast systems, this can be a more significant activity, as millions of users can be involved. Such operators go to great lengths to make the actual cost of billing per subscriber as low as possible, hopefully without confusing or inconveniencing the subscriber.

Customer Service

Once again, this varies significantly with services. For systems leasing capacity, the critical function is often the short-term leases, as they are often used for televising events of topical importance and any outages or confusion can result in considerable customer annoyance or worse.

For direct broadcast systems, customer service can often make a difference in the relative success of competitors otherwise providing equivalent services. For both television and radio broadcast systems, careful coordination is necessary in customer services between the system operator, receiver retailer, and often the receiver manufacturer as well.

QoS and IBSP Feedback

A well-managed company looks beyond short-term success in order to consolidate its market position and return on investment on a long-term basis. While newer technologies and system concepts certainly play a role here, an important element can be ongoing feedback from customer service, as shown in Figure 8.1. Such feedback can point out any repetitive irritant that customers are reporting in the current services; it can also provide useful perspective on what the customers may like to see in the future.

Summary

System operations is critical to the ongoing success of an enterprise in multiple ways. First, it ensures that all of the assets forming part of the infrastructure are functioning normally with minimal risks. Depending on the nature of the enterprise, this can include user equipment as well. Second, operations is the interface with the users on day-to-day basis and thus can influence either way the reputation of the enterprise and its products and services. Third, it provides a valuable feedback on what the market as a whole wishes to see in the near, medium, and often the long term as well.

All of the facilities required for operations should be an integral part of the infrastructure in all respects and right from the start. If some of its functions are to be contracted out, that aspect should also be handled through all of the normal processes of the project. The training of the staff for the infrastructure as well as customer service functions should receive the highest priority and should not be subjected to cost-cutting pressures not uncommon in the early years of an enterprise.

Finally, the maintenance of the necessary QoS is what, in the final analysis, often distinguishes competitors within a given medium and across different media. Operations should lead this activity with all functions supporting it as well as benefiting from it.

References

[1] Snyder, J., and S. Patsiokas, "XM Satellite Radio—Satellite Technology Meets a Real Market," *22nd AIAA International Communications Satellite Systems Conference & Exhibit 2004*, paper 2004-3227.

[2] Verhulst, D., and X. Denix, "DSL in the Sky for Enterprises and Residentials," *Alcatel Space Presentation at International Communications & Satellite Expo (ISCe) 2003*, Long Beach, CA, August 2003.

[3] Sarraf, J., "The Spaceway System: A Service Provider's Perspective," *IEE Seminar on Broadband Satellite: The Critical Success Factors—Technology, Services, and Markets*, October 17, 2000, pp. 15/1–15/6.

[4] Kota, S., and M. Marchese, "Quality of Service for Satellite IP Networks: A Survey," *Int. J. Satellite Communication Networks*, Vol. 21, 2003, pp. 303–349.

9

Managing for Success

> If the Earth could be made to rotate twice as fast, managers would get twice as much done; however if the Earth could be made to turn twenty times as fast, everyone else would get twice as much done since all managers will fly off.
>
> Law #11, Norman Augustine, ex-chairman, Lockheed Martin [1]

The IBSP developed in the preceding six chapters is designed to make the completed project and its deliverables responsive to the starting vision/mission through an organizational paradigm that encourages and facilitates active communication and consistency between different functions. However, this process does not by itself necessarily ensure success. An important element impacting the entire process and even beyond is how good the managerial talent and skills are in the team.

Good management skills are important at all levels of the organization and activities. Such skills become particularly critical at the top for a new enterprise and for an existing enterprise going through a significant change. Correct and timely decisions, or lack thereof, made at the top have much more significant bearing on the overall progress and eventual success of the mission. We will start by looking at such positions.

Visionaries, Leaders, and Managers

As we have noted before, the starting point of most potentially successful enterprises is a clear vision. Very often a single person or a very small group of individuals are behind such a vision. Such visionaries are generally intensely passionate personalities; passionate about the ideas, their importance to society, and their importance to the business world. They work extremely hard—often subordinating their own needs and even risking everything they have—to make their vision a reality. Visions that have clarity and simplicity have the maximum total impact. We have already noted in Chapter 5 Dr. Martin Luther King's "I Have a Dream" speech as one of the clearest and most dramatic vision statements in recent history.

A clearly defined mission statement is just the first, though an extremely important, step to ultimate success. However, a visionary is not necessarily the best person to manage the implementation of the vision, even though he has literally given birth to it and has more intimate knowledge of its origin than perhaps anybody else. In practically every field—politics, science, engineering, and even sports—there are examples of some of the best visions having gone astray or failing mainly because the visionary insisted on leading the detailed implementation himself. This is not at all a reflection on the extraordinary qualities of the visionary as a superb and intelligent human being; rather, it is a recognition of the fact that the skill sets required for building an organization and leading it to success are quite different. Conceiving a vision requires an extraordinary sense of perception for things and sentiments around you, often a very high level of love and caring for humanity, and a talent for a clear definition of what needs to be achieved by a society, country, or company, as the case may be. However, making a vision a reality, through an effective team and well-conceived projects, requires a different and pragmatic, rather than emotional, approach and a set of skills that are often not present, or at least are not the dominant traits, in a visionary. Most important of the qualities needed to efficiently manage complex projects are leadership and management.

Leadership and Management

It is not uncommon to use the terms *leadership* and *management* interchangeably, particularly in connection with successful individuals and companies. This is in fact quite reasonable when discussing the qualities necessary at the top echelons of a company. However, for maximum effectiveness and for placing the right person at the right place, it is important to recognize at the outset the differences and nuances associated with these two key attributes.

This distinction between leadership and management is obviously not a new subject, and it continues to be a key topic for a large number of studies as well as professional management courses [2, 3]. Every time one thinks that the management profession has finally put its arms around these interrelated qualities, out comes another book or a speech by a recognized leader giving yet another perspective. To a certain extent, this is to be expected in the accelerating pace of change in today's world, as leadership and management skills have to be continually refined in order to keep them in step with newer production and operating regimes brought in by new technologies, newer services, and the almost blazing speed of globalization in certain sectors.

Specifically, what are the attributes we look for in a leader over and above the qualities of an efficient manager?

In principle, once you have a carefully developed project plan, all you have to do is to entrust it to experienced and methodical managers to implement it more or less following the script, so to speak. Life is never so simple or organized. Things often go wrong (the well-known Murphy's Law), customers change requirements midstream, resource availability does not follow a preset plan, and so on. Managers need to make changes as they go along. Leadership involves recognizing and articulating the need to significantly alter the direction and operation of the project, aligning people in the new direction, and motivating them to work together and meet the new objectives. Few people are strong leaders who can motivate yet effectively carry out the day-to-day drudgeries of managing. In essence, *management is about coping with complexity, while leadership is about coping with change* [2, 3].

Jack Welch has been recognized as one of strongest leaders in recent history. Not only did he take his company, General Electric, to previously unimaginable new heights, he established several new paradigms for excellence itself. In a recent article dealing interestingly with criteria for choosing a presidential candidate [4], Jack Welch has distilled his definition of a leader into just four key attributes:

1. Successful leaders have tons of positive *energy*. They can go go go; they love action and relish change.
2. They have the ability to *energize* others—they love people and can inspire them to move mountains when they have to.
3. They have *edge*, the courage to make tough yes-or-no decisions—no maybes.
4. They can *execute*. They get the job done.

Successful leaders at the top generally have strong personal traits. Their life generally revolves around solving problems almost round the clock in order to meet their mission or their own self-imposed goals. However, such strong personalities do often have difficulties in coexisting with each other in the same company—the proverbial *limited room at the top*. A recent example was the merger of two industrial giant companies, Hewlett Packard (HP) and Compaq. At the time of the merger, the head of Compaq, Michael D. Capellas, announced that he was looking forward to working as the president of the combined company, still called HP, reporting to the CEO, Carly Fiorina. While they genuinely got along with each other, the inner drive to be nothing but number one was so strong for Mr. Capellas that within a short time he left the relatively safe haven of a sound and healthy company, HP, to go to WorldCom, a company barely surviving under extremely difficult conditions, buried in scandals and with very limited chances of survival. What made Michael Capellas take such a risky step in mid life? Was it the intense desire to be right at the top or was it the challenge of fixing the unfixable? Perhaps both aspects played a role in his decision to jump out of a secure position. It is often the case with true leaders that they are attracted by almost impossible challenges of bringing about a change even under most unfavorable circumstances. But they want to be in command and are willing to live or die by their actions and decisions. The seemingly impossible challenges with potentially attractive rewards and recognition down the line, notwithstanding their odds, sometimes act as an irresistible elixir of inspiration for really strong leaders. As of early 2004, WorldCom was on the way to becoming MCI and emerging out of bankruptcy.

Good Managers Walk on Water

While some individuals begin to demonstrate managerial (and often leadership as well) qualities fairly early in their professional careers, most professionals do not start as managers. In fact, one of the hardest transitions to make is to move from being a one-person creative member of staff to a section lead or a manager. This is particularly true for those who have had a highly productive start in professional life with a number of *individual* achievements to their credit. When they start as a manager for the first time, they tend to demonstrate a high degree of impatience with their team members. Without saying so, the new manager expects even a rookie new staff member to be as quick, productive, and creative as he himself was just before assuming managerial duties. In his impatience to succeed right away, the new manager often forgets that it took him several years to reach that level, assisted no doubt by the help and patience of his own managers. In extreme situations, the new manager starts doing the work himself, thus further demoralizing his team members. However, with a careful vigilance and polite guidance from his superiors, in most cases the new manager soon begins to appreciate the exponential power of synergy among teammates and gradually begins to derive a different kind of professional satisfaction from the achievements of the team as a whole. What often convinces the young manager is that there is a long-term future in bigger and bigger *group* achievements compared with the real risk of being rendered obsolete in *individual* achievements by the never-ending supply of younger and brighter members of technical staff (or equivalent). Such a transition to managerial responsibilities is not unique to one field or another; rather, it needs to be watched for in practically all disciplines and even at higher strata of organizations.

Project managers at junior and mid-levels are generally selected from within the organization from the pertinent line functions. The selected person should have credibility, sensitivity, and a strong sense of ethics. Good technical background is important, not necessarily for personal technical contributions but for the ability to ask pertinent questions to specialists and to select the best course or solution from several alternatives. Some of the qualities in an effective manager are [2, 3]:

- Be a systems thinker;
- Have personal integrity;
- Have a high tolerance of stress;
- Have a general business perspective;
- Be a good communicator;
- Demonstrate effective time management;
- Be a skillful politician;
- Be an optimist.

Of course, life is not always fair and a manager can be buffeted with contradictory pressures such as:

- Be innovative yet and maintain stability;
- See the big picture while getting hands dirty;
- Encourage individuals but stress the team;
- Be hands off yet hands on;
- Be flexible but firm;
- Have team and organizational loyalties.

Network of Relationships

The old saying "no man is an island" also applies to most managers in the modern world. Even in a line or functional organization, a typical manager has to interact with her peers for various inputs as well as support in various forms. First-time project managers soon find out that their abilities to complete their responsibilities depends not just on the designated project team but also on support from others, sometimes to a much larger extent than they anticipated.

There are different categories of such *project stakeholders,* who need to be managed through what is frequently referred to as a *network of relationships.*

Figure 9.1 shows such a generic chart [2]. In this example, the project is *embedded* in a bigger organization structured largely on function lines but with several ongoing projects.

The project manager tries his or her best to manage the project with the allocated resources. However, the dynamic nature of the project often requires support from other project managers for manpower, facilities, and testing. In certain cases, he or she may need fresh approvals from senior management for additional resources.

The *customer* can be the ultimate users of the system or a service provider leasing capacity on the company's space segment. In the latter case, good ongoing relations can play a critical role in smoothing out minor problems of scheduling meetings and reviews, for example. At the other end, there can be multiple contracts with the industry for the space segment, ground network, content facilities, and business systems. Once again, good relationships, while not relaxing contractual requirements, can go a long way in a smoother progress on all deliverables.

Relations with the investors may or may not be necessary in all cases; in most cases, such relations get managed through senior management. Along the same lines, relations with the regulators can be routine for well-established

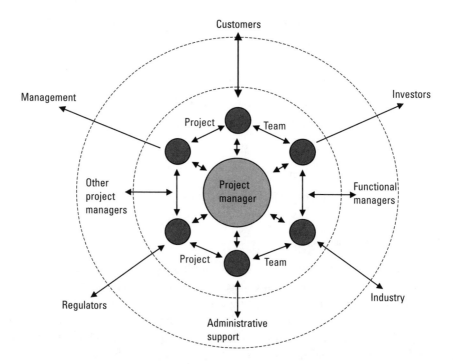

Figure 9.1 Network of relationships. (*From:* [2]. © 2003 McGraw-Hill, Inc. Reprinted with permission.)

frequency bands but could require a fairly active dialogue in the case of totally new bands and services.

In such relationships within the organizations and with other project managers, the principal of give and take does apply. In other words, the manager has to be ready to use some of her common *tradable currencies*, such as staff resources and testing facilities. Within her team, a good manager can sometimes use what are known as *inspiration-related currencies*. These include explaining the overall vision and the importance of excellence in personal development in the organization. While using any such currencies, at no stage should there be any relaxation of ethical standards [2].

Tools for Success

As the art of management in general and project management in particular progresses, better and newer tools continue to become available for enhancing team efficiencies and their chances for success. Many such tools emerge from large government programs and have been extensively reported [2, 3]. Others continue to improve in their power, scope, and flexibility thanks to rapid advances in computer technology. In this book, we have developed in some depth the IBSP. In this section, we will present only a few of such tools that have a real potential for benefiting satellite-based and similar programs. We begin with the benefits from the IBSP in terms of substantially enhanced teamwork and consistency through interactions.

Active Interactions and Not Walls

In the previous chapters, specific groups of activities in the IBSP have been addressed in turn. In each case, applicable interactions with other groups have been highlighted. We will now take a combined overview of all of the major interactions and highlight their importance and benefits. Figure 9.2 shows the IBSP of Figure 3.1, with only the most significant interaction pathways added for the sake of clarity. In actual practice, several additional temporary and permanent pathways develop often on their own once the working climate is conducive to encouraging teamwork and interaction.

In Figure 9.2, there is an implicit assumption that all of the functions exist at the time this specific project is initiated. In other words, it assumes that the project belongs to an already operating company or organization. If it is a new startup, it is expected that the founding members of the project or company would bring in the requisite expertise through outside support until the funding permits them to start manning the organization. We will now highlight and recall the importance of some of the forward and backward interactions and the benefits that would accrue in each case.

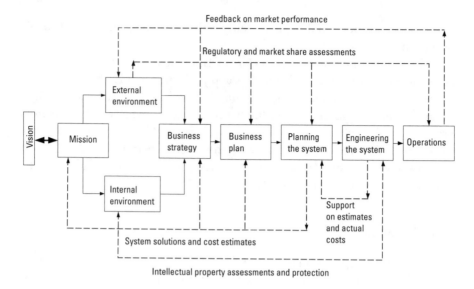

Figure 9.2 Examples of interactions through the IBSP.

The first major stage of interactivity is at the time of development of the business strategy, when viable strategic choices responding to the mission are developed, as discussed in detail in Chapter 4. Through the associated external environment analyses, the strategy team surveys the competitive landscape and identifies the market share that would form the basis of the enterprise. A critical next step at this stage is the identification of the appropriate system architecture and infrastructure and its costs. Without such inputs, the strategic choices made at this stage may carry with them considerable risks, some of them latent and possibly serious.

In a structure with little interaction and relative isolation (e.g., the stove-pipe model), the strategy team members may try on their own to make rough estimates for costs and schedules *by similarities* with other systems—perhaps by merely *surfing* and "Google-ing" the Internet. Except for simple ventures or projects almost totally mirroring an established competitor, the answers can be incomplete or even dangerously risky. A much better approach is to involve IBSP groups, such as system planning, supported by the engineering teams. Even top-level inputs on the likely architectures and costs from such groups can make the process of strategy development more accurate and often make the convergence process to the final strategy more authentic. Of course, such inter-active steps generally also set the stage for a more efficient system against competition.

Much of what was said earlier about the business strategy activities also applies with even greater force to the development of the business plan discussed in Chapter 5. Unlike an internal strategy document, the business plan document and its associated presentations are presented to potential investors and partners, and therefore their quality and authenticity can determine the very future of the enterprise. Despite this obvious realization, the practice of "business plan on a napkin" in a restaurant still survives from the go-go days of the dot-com era, and bright strategists and future CEOs continue to believe that the intensity of their convictions in their project is enough to convince would-be investors for the long haul. While a solid discussion across a table in a café with potential investors *who are familiar with your background and achievements* may still be enough to obtain seed money to get started, there is no substitute for a well-balanced and credible business plan for long-term investments from the potential owners and the public at large.

As Chapter 5 underscores, a business plan is in many ways a microcosm of the organization, and it should be as accurate as possible in all of its components. The strategy should be based on environmental analyses that can withstand scrutiny; the costs should be as accurate as can be assessed by the industry; and the schedules should be plausible and defensible. The planned products and services should match what the market data says are needed and not those that you may wish to thrust on the market. And, finally, the risk assessments should be as accurate and open as possible. They should cover all aspects of the project, extending from market data, potential regulatory changes or uncertainties, technological challenges, and vagaries of the insurance market. As we have discussed in several places, there are also hidden risks in ignoring the time sensitivity of the market data, and therefore the impact of significant delays in the project implementation should be estimated and highlighted. It is obvious that business plans meeting these diverse qualities and attributes can only be developed via a very high degree of interaction with practically the entire organization.

Moving down the line of the IBSP, perhaps the most critical interaction is not so obvious from Figure 9.2. This is the interaction between the two adjacent boxes, planning the system and engineering the system. While the IBSP deliberately shows these functions next to each other, in many organizations what we have called *system planning* in Chapter 6 is not an explicit activity but rather is spread in some fashion across multiple units, such as business planning, project office, or strategic planning.

The system-planning group should be perhaps the most interactive of all. As we have seen before, its potential support can commence right from the stage when the founding group gets started. By their nature and charter, system-planning teams have the talent to convert diverse and rather loose market assessment into potentially tangible operational concepts. An equally important value of such a group is in supporting the engineering group in converting the marketing information into system requirements for a realizable infrastructure. As we have seen in Chapter 6, this process starts with a review of alternative approaches that get narrowed down through a design-optimization process on the basis of major criteria such as costs, schedules, risks, and acceptability in the marketplace today and tomorrow.

Another interactive role discussed in Chapter 7 deserves to be highlighted here. Once the engineering and contracting process gets underway, there are different pressures and temptations to add new features to the system. These are motivated either by technological advances or by the interests of the industry or both. The close coupling between system planning and engineering on one hand, and system planning and the strategic and business groups on the other, can ensure that any changes that are made in the engineering objectives are within the overall business strategy and with the marketplace. In summary, the system-planning group can play two main interactive roles:

- Act as a central point for translating both ways the language of the market into the language of the engineers—specifically, provide data to all other functions what can be achieved by the engineering state of the industry and at what cost and time frame;

- Act as an interface and a watch dog of engineering in terms of keeping the infrastructure and its costs to what is needed and not what is possible.

To a certain extent, we have covered in Chapters 6 and 7 some of the interactivity as it relates to the engineering group. There are a few additional and important roles that should also be noted. While handing over the developed system to operations, engineering has to make sure that the operatives are properly trained and all documentation is complete. This is particularly critical in the case of satellite operational networks discussed in Chapter 8. Given the 15-year-plus life of modern satellites and the continuing changes in the industrial world, it can frequently happen that halfway through the operational life of the satellites, the original manufacturers no longer exist or, as is most likely, the industrial staff with the intimate knowledge of details has moved on and is therefore no longer accessible. This makes it extremely critical that the engineering groups gather all documentation and test data throughout the process and hand over the same to the operation teams. While it may not be true in very large organizations, it is a fact of life that once a project is implemented, its own engineers also move on to different careers or organizations.

In sum, the primary value of the IBSP is in ensuring and encouraging interaction between different groups. An organization with walls between teams, however competent in their domain, cannot achieve as much success as those who adopt organization structure and culture akin to the IBSP.

Progress and Performance Evaluation

> Every time you review a project, it slips by one-third and its cost goes up by one-third...therefore the best approach is not to hold progress reviews at all.
>
> —Norman Augustine

Norman Augustine, former Chairman of Lockheed Martin wrote a classic book a few years ago, documenting in a humorous manner serious advice for managers at all levels [1]. We have quoted one such law at the beginning of this chapter. This second quote once again uses humor to drive home the point that managers should watch out for delays and cost overruns as the project progresses through external contractors. In real life, no manager can of course afford not to monitor the progress and performance of her project teams and contractors at periodic intervals. Such monitoring is essential, not only for reporting purposes, but also for making mid-course corrections if needed in a timely manner.

Recently, in my class on project management, I asked the students if they already use project management tools such as Gantt charts. The answers were partially surprising. Full-time students and those with a natural flair for modern software had already discovered on their own some of the conveniences of modern software systems as an aide for managing projects. However, several students who were members of project teams in their regular jobs gave evasive replies at first. Finally, one of them came forward to say, "...only when presentations have to be prepared for some visiting bigwigs!" After some more discussion, it transpired that most of their projects were being run quite well; however, the reluctance to use modern tools such as Gantt and other charts was more of a carryover from the times when it took too much effort to make such charts that invariably became obsolete even before they could be drafted, duplicated, and distributed. However, this is no longer the case, and modern computerized tools to assist project managers are much more versatile, easier to use by the whole team on a real-time basis, and can provide powerful insight at all stages of the project. For example, the current versions of programs such as Microsoft Project or equivalent can:

1. Provide a consistent framework for planning, scheduling, and monitoring a project, however complex. Such programs in most cases can be accessed by all team members in real time and can be a powerful tool for review meetings in conference with participants at different locations.

2. Develop clear identification of the interdependency of different tasks, work packages, and work elements. Work breakdown structure (WBS) is part of such programs and is automatically created and updated.

3. For each element of the WBS, the resources can be defined and collated right from the start. These background databases include both types of resources, facilities and staff, with their costs and schedule sequencing.

4. Determine and update slack times for each task and intermediate and final milestones. Identify critical paths through matching critical path model/program evaluation and review technique (CPM/PERT) displays at the beginning and throughout the program.

These are only top-level benefits of simpler programs. While at the industrial level, custom programs have been in use for quite some time, at the project or customer levels, even otherwise computer-savvy executives tend to shun such tools, lest they be labeled as "techies" rather than management types. However, without good tools, the managers at all levels run the risk of not getting the latest updated information and a quantitative assessment of its impact in terms of schedule, resources conflict, and changing critical paths.

Integrated Cost/Schedule Monitoring Systems

In most projects of even moderate complexity, one can draw incorrect conclusions by looking in isolation either at costs or schedules at any of the intermediate milestones. For instance, expenditures in excess of the planned amount at a given milestone may not necessarily signify an over run, as it is possible that more work has been completed than planned. Conversely, a lower level of costs does not always mean that there are real cost savings. For a meaningful assessment, it is necessary that both costs and schedules are reviewed in an integrated fashion against the nominal plan. Such assessments are generally carried out through *earned value charts* [2, 3], an example of which is shown in Figure 9.3 along with a glossary of terms used.

The usefulness of such integrated monitoring systems presupposes that a baseline cost versus profile (labeled BCWS) has been developed after a reasonably accurate WBS and network analyses. At the time of mid-course assessment, the BCWP plots the budgeted cost of work performed, while another report captures the actual cost of work performed (ACWP). In Figure 9.3, the BCWP graph is below the nominal while the ACWP is above. In this particular case, not only has less value been realized, it has been done at a higher cost. Therefore, there is a spending variance as well as a schedule variance, as shown on the graph on the right. It should be noted that schedule variance is measured in dollars, while the value variance is measured in terms of time; these concepts can take certain effort to getting used to.

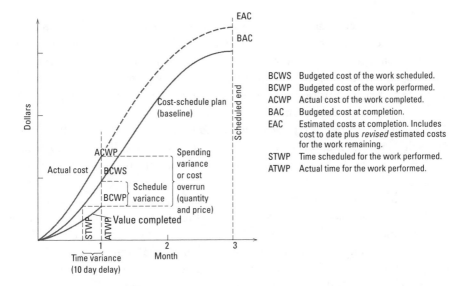

Figure 9.3 Earned value chart example. (*From:* [3]. © 2003 John Wiley & Sons, Inc. Reprinted with permission.)

The earned value chart approach is useful not only for customer organizations but also for different contractors for their own internal program management and controls.

Managing Virtual Organizations

Modern text books document extensively three classical types of organizational structures:

- Functional organizations;
- Project organizations;
- Matrix organizations.

The three categories are largely self explanatory and have their own relative advantages and disadvantages [2, 3]. Most organizations start off as a functional organization, introduce some project teams, and, if needed, move on to a matrix environment in case resources have to be shared among several projects. The IBSP can be considered either a functional organization or a project organization, depending on how it is viewed and structured. Perhaps the best way to characterize it would be as a *project team with a highly interactive functional structure.*

For satellite-based systems and in several other fields, more and more real-life scenarios need something more outside the IBSP as well. They go beyond traditional contracting and partnerships. What is often needed is a flexible structure going across company, and often national, boundaries as well, bringing together many enterprises committed to a common project goal and schedule. Such *virtual* organizational structures are a result of expertise and cost effectiveness breaking across traditional geographical boundaries and the globalization of products and services in general.

Figure 9.4 shows a typical virtual organization that is behind the recent successful implementation of modern DTU satellite-based broadcast systems. With some variations, these could apply to a system for broadcasting television or radio.

At the center of the chart is the project office, with the responsibility to implement the whole project and set the stage for successful and efficient business operations. On the right-hand side are grouped entities that are related to the consumer equipment and its distribution. The OEM industry can be the car industry for digital radio, as an example. In the lower left-hand side is the content-generation value chain with both in-house and external elements. The upper right shows the satellite-related activities.

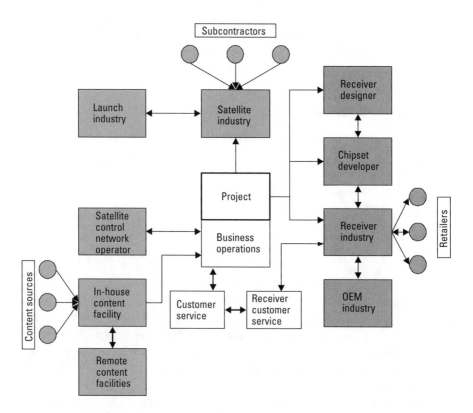

Figure 9.4 Typical virtual project organization.

At the bottom is customer service, both for the system as well as for the consumer equipment, hopefully fully coordinated.

At first glance, one could interpret this chart as a normal contracting chart for a central organization. However, for maximum efficiency, cost effectiveness, and timely coordination of all of the diverse actions in a fully coordinated fashion, modern project-management tools need to be applied to the whole virtual organization over and above any contractual requirements.

And Finally, How Do We Measure Success?

Satellite technology is now almost half a century old, during which period literally hundreds of commercial, scientific, and government programs have been implemented. However, the excitement associated with the rocket shooting toward the skies carrying the latest payload has only partially abated. To a certain extent, this continued exuberance is understandable, as not only is a rocket launch still a pretty exciting event, but a successful launch does represent a major risk reduction financially. Unfortunately, often such exuberance does not just stop there; it is sometimes taken as a major measure of success for the overall project as well.

During the period that almost all of the satellites were meant for transponder leases or telecommunication links, a launch of a satellite did represent a good measure of success because in most cases the successfully launched satellite would soon add more capable transponders to the then existing pool. However, with the increasing emphasis on DTU services, the true measures of success have to be related to the overall business rather than only to the completion of an important, albeit exciting, milestone.

An objective set of success criteria is generally made up of some hard quantifiable criteria and some soft criteria. We will lean on previous chapters to address both such groups of criteria.

In Chapter 7 on engineering the system, we developed a few measures of success that also constitute the hard criteria for the overall project as well. These are recalled here with slightly broader definitions:

1. Did the project meet, without hidden subsidies, the target prices charged to the consumer for both the equipment and subscription charges?
2. Was the project completed on schedule?
3. Does the final system have adequate flexibility and fungibility? (This is also in the following soft criteria.)
4. Did the project meet the financial projections and criteria committed in the business plan?

For illustrating the soft criteria, we turn to Chapter 2 and recall the list of lessons derived from recent satellite-based projects:

1. *If you are in consumer business, start with the consumer.* This generic lesson underscores the importance of focusing above all on the needs of the consumer and his costs. In other words, was the project managed as a consumer-centric one or a satellite-centric one?

2. *Reduce the lead times as much as possible or your business plan will become somebody else's.* In a landscape of several competing alternative media and newer technologies emerging all of the time, market projections are becoming more and more time sensitive. A significant schedule slippage would often allow an alternative system to take away the targeted market share. Was the project completed on time?

3. *Unless you have positive solid confirmation of the business over the satellites' lifetimes, avoid making nonfungible satellites.* This criterion, as we recognized earlier, is a hard one and is becoming harder as newer services and technologies often mandate a space segment that is highly application specific, with very little fungibility to do something else efficiently. This is a risk that should be taken if there are tangible benefits to be derived *and* there is good and verifiable data available to the management that the target market share is credible and the schedules will be met.

4. *A system capable of growing in a modular fashion in the space segment has greater ability for mid-term corrections.* This continues to be true and should in fact be given even higher importance in today's uncertain markets.

5. *Underserved market segment are often also financially untenable.* Again, this goes to the heart of market-share determination. Competing head on with alternative media in the core market is harder, but that is the only realistic test of competitiveness.

6. *A successful business model in one part of the world can fail elsewhere.* Despite rapid globalization, the world's economies are still quite diverse in terms of relative purchasing powers and cultural preferences when it comes to content.

7. *System dynamics of a two-way system is quite different from that of a one-way system.* This is a good checkpoint before one jumps into seemingly similar two-way markets.

8. *In good times and bad times, boring is often good.* This is a good antidote to the "gee-whiz" aura long associated with space technology. An

objective check should be made to see if the system developed has bells and whistles that cost a bundle but did not add to either revenues or customer appeal.

References

[1]   Augustine, N., *Augustine's Laws,* 6th ed., Reston, VA: American Institute of Aeronautics and Astronautics, 1997.

[2]   Gray, C., and E. Larson, *Project Management: The Managerial Process,* 2nd ed., New York: McGraw-Hill, 2003.

[3]   Meredith, J., and S. Mantel, *Project Management: A Managerial Approach,* 5th ed., New York: John Wiley & Sons, 2003.

[4]   Welch, J., " Four E's (A Jolly Good Fellow)," *Wall Street Journal,* January 23, 2004.

10

Digital Radio Systems Case Studies

Extra-Terrestrial Relays: Can Rocket Stations Give Worldwide Radio Coverage?
—Arthur C. Clarke, October 1945

This quote is the title of the landmark paper that a little over half a century ago launched the whole field of satellite communications [1]. In the opening paragraph of the paper, Clarke made a prescient forecast, "A true broadcast service...will be invaluable, not to say indispensable, in a world society." Based on some contemporary developments at that time, Clarke's estimate was that such a project might take 50 to 100 years. Well, it did take a little over 50 years to realize the first satellite-based digital radio system. However, the distribution of radio and television programs to local stations via satellites started as early as the late 1960s.

A few years ago a young man was driving his car deep in the interior of Central Africa. His objective was to test a newly developed portable satellite radio in a mobile environment. To carry out such a test, he had mounted the detachable radio antenna of his radio receiver on the roof of his car with a few pieces of Velcro. The receiver itself was on the passenger seat by his side with the antenna cable snaking through the rolled-down window on that hot summer day. Around noon, he stopped at a small hamlet to get some cold water, both for himself and for the engine of his car. As he was parking his car, he was all of a sudden surrounded by tens of people of all ages, all with their wide-open and curious eyes and some even with a hesitant smile on their lips. After exchanging a couple of broken sentences in the local language, he learned that the villagers were clustering around him and his car not just because he had an innovative new radio working directly from a satellite. In fact, they were there for a more

157

fundamental reason: They were highly curious to find an explanation for what they had never encountered before in that remote hamlet in the center of Africa. They were indeed enchanted by the melodious music surrounding them, but they could not find any singers in the car or even any musical instruments! It took the city-born young man a few minutes to comprehend that *it was the first time ever that the people in that hamlet were listening to any kind of recorded or broadcast music!* This was because there were absolutely no radio stations any-where near, and of course none of the traders in that little village sold anything like recorded music. If there were any doubt in his mind about his new job, it disappeared after that day. You see, he had recently joined the world's first satellite radio company, WorldSpace; the year was 1998, and Africa was the first region in which they had started beaming music and information from all over the world.

This chapter examines the case histories of the several digital radio systems in operation today and those under planning. We will do so in the context of the business strategy principles developed in this book.

Introduction to Digital Radio Systems

For most of the first half of the twentieth century, radio technology evolved on roughly two concurrent streams. One was the broadcast of programs beyond national borders. For this purpose, the vast distances involved were bridged via HF radio systems, connecting for the first time totally diverse societies through live news and music. Most of these systems were government run and often became synonymous with propaganda. The market response was therefore mixed; however, they did achieve reasonable popularity because a good part of the information broadcast from distant and often more developed societies was often quite useful to daily life. The program choices were small, and their quality was often spotty and undependable, but in the absence of something better, such services sustained over decades and still continue in several parts of the world. Despite their limitations, such broadcasts did and still continue to provide societal benefits in the fields of education, medicine, and agriculture.

The second stream evolved around local AM/FM radio broadcasts, both at home and in automobiles. While the home radios soon had competition from television, the radios in cars had access to no other media alternative, and this mode has grown literally on the back of diverse transportation vehicles throughout the world. In 2002, there were nearly 2.5 billion radios worldwide, of which over 750 million were in cars and trucks.

By the early 1980s, the satellite technology had advanced sufficiently to create serious system interests in direct broadcast systems. While the bulk of the focus was for television due to its far larger established business potential, there was keen interest in different forms and in different parts of the world to harness this technology for the good old radio as well. Progressively, three separate and partially overlapping visions emerged:

- Replacement of the HF broadcasts from the developed countries to the developing world with satellite broadcasts of superior quality and with larger programming choices;

- Direct broadcasts to automobiles with much larger coverages than AM/FM stations and greater choices;

- Use of satellites to provide the world's poor and developing nations with much-needed information for education and for accelerating the pace of their economic development.

Each of these visions has followed historically interesting and somewhat unexpected paths to fruition in different parts of the world.

The operators of the largely government-owned HF systems in Europe and the United States were in fact first to actively evaluate the alternative system concepts [2] and made some important contributions in international forums, including the ITU [3]. While they had serious interest in building satellite systems at frequencies much higher than HF, they never could get adequate funding, given the perennially uncertain budgetary allocations for such services.

Radio broadcast to cars was eventually recognized as a much larger and more lucrative market; however, it also ran into some obstacles of its own for quite some time. In Europe, the technology for terrestrial radio broadcasting was very actively pursued; however, the large regional bodies of broadcasters and PTTs could not come to a common technical approach for using satellites and terrestrial transmitters in a complementary fashion. In the United States, while there was no similar agnostic attachment to specific technologies, the powerful organization of small independently owned local broadcasters, the National Association of Broadcasters (NAB), remained fiercely opposed to any nationwide direct broadcast satellite system. Eventually, as we shall see in more detail later, the United States did build two systems to cater primarily to this segment.

The third type of vision for using satellite broadcasting to alleviate the scarcity of information for the betterment of the world's poor and developing nations was surprisingly successful relatively quickly and had the privilege of in fact building the world's first digital satellite system. As we shall see in greater detail, it had to tackle all kinds of novel challenges.

Before we look at specific case histories, it is helpful to highlight two issues common to all systems: spectrum and compression technologies.

Spectrum

The first hurdle for any new type of radio system is of course availability of spectrum. Due to the lukewarm business interest in this arena, it was not at all an easy task for the early pioneers to get any type of spectrum allocation agreement at the ITU. After considerable efforts, the World Administrative Radio Conference, Malaga-Torremolinos, in 1992 recommended the allocations shown in Figure 10.1.

All of the three segments shown in Figure 10.1 are now in use in different systems around the world. As we shall see in more detail in later sections, the L-band spectrum has been, and continues to be in certain parts of the world, a competing ground between the protagonists of the terrestrial and satellite approach to digital radio. While such tussles are not in any way new to the spectrum world, they assume a much greater importance when we realize that the total intrinsic capacity of these longer wavelength bands is quite small, particularly in the context of the paramount market need to work with inexpensive nontracking small user antennas. We shall revert back to this aspect when we review the plans for a hybrid system for Europe.

As Figure 10.1 shows, there are two segments in the S-band allocated to digital audio broadcasting. The lower segment, 2,310 to 2,360 MHz, is currently in use in the United States, but only just over half of the total 50-MHz bandwidth is recommended for this service. The FCC in its order granting the two licenses in March 1997 [4], decided to allocate the remaining 25 MHz to other services within the United States. The upper segment in the S-band was being used in the MBSat satellite system launched in early 2004 for service over Japan and Korea.

Compression Technologies

As will become apparent as we go into the case histories, three major technologies have enabled the realization of direct broadcast systems for both audio and

BAND	COMMENTS	UTILIZATION (2003)
1,452–1,492 MHz (L-band)	Digital audio broadcasting on a shared basis in all countries except United States Eastern Europe, Russia, and India	Europe terrestrial systems, WorldSpace (Africa, Asia); Europe hybrid system (under planning)
2,310–2,360 MHz (S-band)	United States; The FCC limited the available spectrum to 2,320–2,345 MHz	Sirius(United States); XM (United States)
2,535–2,655 MHz	Eastern Europe, Russia	Japan (MBSAT, 2004)
2,535–2,655 MHz 2,310–2,360 MHz	India, Japan	

Figure 10.1 Digital audio broadcasting spectrum allocations.

video. Compression technology is one of them, and the other two are satellite power and the availability of reliable and affordable ASIC chipsets for consumer radios for such services.

Without compression of audio signals at the source, the overall capacity of digital radio systems in the available RF spectrum in terms of choice of programming will be very small and well below the threshold of the economic viability of such systems. Accordingly, realization of better codecs for higher level of choices and quality is one of the continuing drivers for practically all digital broadcast systems for audio as well as video. Each successive generation of compression codecs needs less baseband bit rate for the same subjective quality. In most cases, lower bit rates also translate to lower RF spectrum needs, unless, of course, greater information channel choices are provided. The compression technology has advanced almost concurrently in time as the different systems have been realized in different parts of the world [5, 6]. The impact of this factor can be seen in the spectrum occupied by different systems implemented at different time frames. Figure 10.2 shows a summary of subjective test results obtained in 1998 with the different compression techniques available at that time [7].

The progress here has been quite rapid and shows no signs of saturation, given the ongoing progress in the underlying computing technologies. The compression technology for audio has advanced today to such an extent that digital audio occupies less bandwidth than an analog signal for the same quality and much lower bandwidth than the compact disc standard. Current ISO MPEG 2 audio layer 3 (MP3) provides quasi-stereo compact disc quality at 64 Kbps, which is reduced by more than 15 times compared to compact discs. New ISO MPEG-2 AAC/AAC+ audio coding provides compact disc quality at 48 Kbps [8].

As a backdrop to the case histories that now follow, Figure 10.3 shows the regions in which the different digital radio systems are either already in operation or under planning. Table 10.1 summarizes the very top-level features of these systems.

The following sections provide more detail on each of the systems listed in Table 10.1. It should be noted that excellent descriptions for many of these systems have been published by the ITU [9].

Europe (Eureka 147)

In many ways, Europe has played a leadership role in the realization of digital radio systems worldwide. Not only did this region pioneer the related system concepts and technology development, but its industry has also developed and provided several major components for such systems for other regions.

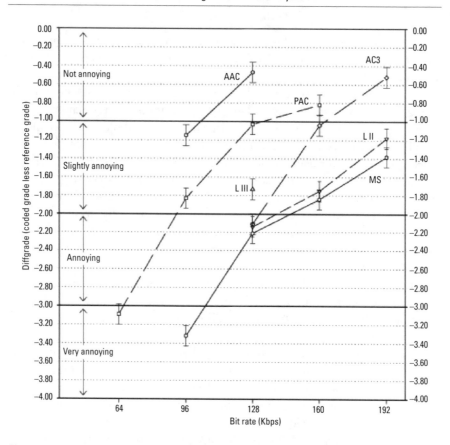

Figure 10.2 Subjective quality tests (1997). (*After:* [7].)

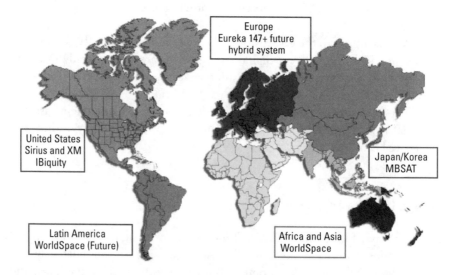

Figure 10.3 Geographical distribution of digital radio systems.

Table 10.1
Major Digital Radio Systems (2004)

	Europe	WorldSpace	XM Radio	Sirius Radio	IBOC	MBSat	Europe (future)
Architecture	Terrestrial only	Satellite + Terrestrial	Satellite + Terrestrial	Satellite + Terrestrial	Terrestrial	Satellite + Terrestrial	Satellite + Terrestrial
Operations Start	1995	1998	2001	2002	2004(?)	2004	?
Mission	Higher quality broadcast to current AM/FM listeners	Information and entertainment broadcasts in developing areas	High-quality nationwide alternative for cars	High-quality nationwide alternative for cars	Digital technology for terrestrial AM/FM transmitters	High-quality audio, video and data broadcast	Likely a combination of U.S. and Japanese missions
Target Market	Europe	Africa, Asia, Latin America	United States	United States	United States	Japan and Korea	Europe
Spectrum	Lower part of L-band allocation	L-band allocation	Lower S-band 12.5 MHz	Lower S-band 12.5 MHz adjacent to XM	Within current AM/FM allocations and constraints	Upper S-band 25 MHz	Possibly full L-band subject to coordination with WorldSpace
System Configuration	Single-frequency networks at L-band other bands	Three-beam spacecraft design common to three regions	Two GEO spacecraft with time/space diversity + terrestrial rptrs.	Three inclined orbit spacecraft with time/space diversity + terrestrial rptrs.	Add-on equipment to existing transmitters; new consumer radios	One GEO satellite with a wide range of gap-fillers	Could be similar to XM or Sirius
Future Evolution	See last column	Mobile operations in selected areas	Next generation satellites with enhanced services	Next generation satellites with enhanced services	Licensing of current AM/FM broadcasters		

The driving vision for the very first European efforts in the early 1990s was to develop and exploit digital technology for much higher quality of broadcast audio than currently available to existing listeners of AM and FM radio, both at homes as well in the mobile environment [10]. It was quickly concluded that the existing frequency bands were inadequate for the needed bandwidth, and this conclusion was measurably instrumental in the eventual ITU allocations shown in Figure 10.1. However, instead of a regionwide new system, a series of terrestrial overlay L-band networks, generally congruent with the geographical reach of the existing AM/FM systems, were envisioned. This factor was partly instrumental in relegating the use of satellites to some future date.

In order to meet the dual objectives of enhanced quality and mobility, a series of well-coordinated development programs were launched under the generally umbrella of what came to be known as the Eureka 147 program [11]. The overall management was through different levels of intra-Europe groups involving the national and regional broadcasters, the European Broadcasting Union (EBU), the PTT administrations, and several excellent state and private industrial organizations.

Given such a framework, the introduction of digital radio systems was not a self-contained business venture but rather part of a long-established and successful process of introduction of newer technologies through various national agencies.

Looking back, perhaps the greatest contribution of the early efforts in Europe was to establish the merits of coded orthogonal frequency division modulation (COFDM) as a practical means of counteracting multipath effects and in the realization of single frequency networks [10, 12, 13]. Their unique feature wherein all of the terrestrial repeaters can radiate at the same RF frequency has made digital radio broadcast practical in a terrestrial environment not only for the European terrestrial systems but also for the terrestrial repeaters of hybrids systems elsewhere.

However, the decision of the organizations involved to stay exclusively with terrestrial infrastructure prevented this technology from receiving widespread acceptance in Europe well ahead of other regions. In particular, as a recent review points out [8], today, even after 10 years, the European Union 147–based terrestrial-digital audio broadcast (T-DAB) has enjoyed only limited development. The main reasons stated in the referenced paper are:

- A limited number of attractive programs compared to the current analog radio offerings, especially in FM;
- A limited coverage, due to the lack of a huge terrestrial infrastructure investment needed in a terrestrial-only solution;
- A limited capacity of programs, due to the use of rather old, inefficient audio coding scheme (older than MP3);

- A limited forward error correction (FEC) performance, optimized only for the specified audio coding scheme but not to higher performance audio coding schemes, nor to data transmission, which needs high error protection.

In summary, the digital radio–related efforts in Europe during the 1990s onwards have been truly pioneering and in many ways paved the way for this technology to blossom in different parts of the world. However, in terms of business-strategy principles, these initiatives have followed an inconsistent approach that has perhaps translated into a lesser degree of success than truly deserved by the level of efforts. The starting vision was clear and precise: provide better quality and choices to radio listeners through digital technology. However, very soon this clear customer-oriented strategy was supplanted by the somewhat parochial interests of the stakeholders—the national and regional administrations. As a result, the progress has been mixed across the continent with different roll-out strategies in different countries. An unintended fallout of this strategy was that an adverse reaction appeared to develop toward the use of the satellites themselves. There were two distinct factors contributing to the reluctance to add a satellite component. The first was the assessment that listeners definitely want a local programming element, and therefore progressive roll out of the terrestrial solutions was the correct approach. The second was a strong desire to use exclusively the European-developed COFDM-based standard throughout the system, which made the satellite component prohibitively costly due to well-known technical factors associated with this technique [9].

WorldSpace

Over the course of history, national and international organizations as well as kind-hearted philanthropists all over the world have regularly come forward to conceive and to implement small and big projects aimed at improving the standard of living of the less fortunate among us through a variety of means, including education, medical assistance, and vocational training. Almost all of them do so either with public money especially appropriated for such purposes or with their self-acquired wealth as a personal gesture of their gratitude to society. They rarely do so through a capital-intensive project funded almost entirely by traditional investment capital. However, that is what a Western-educated lawyer, Noah Samara, still in his thirties and living comfortably in Washington, D.C., set out to do in the early 1990s, after witnessing firsthand the plight and ignorance of the people living in his native continent of Africa.

That was not the only difference in his approach. Instead of distributing funds to local welfare institutions and hospitals, Mr. Samara decided to go to the root of the problem and, after extensive on-the-spot reviews, came back convinced that widespread poverty and devastating diseases like AIDS were not because of "any fundamental DNA differences" [14] between successful rich healthy people and poor deprived people. It was instead substantially due to lack of proper education and information about the latest in technology and health-care in vast areas of our civilization even in the twentieth and now in the twenty-first century. For more than a decade, he has relentlessly and eloquently spread his conviction that, "information is the *sine-que-non* to social and economic development" [14].

The solution adopted by the newly founded company, WorldSpace, to alleviate rampant information scarcity spread over vast regions was to broadcast the much-needed information directly to the target population through powerful satellite systems. The "messenger" to deliver the information to the millions of ultimate users could be either television or radio. The technology for direct television broadcasts was already maturing in the West, but was ruled out due to extremely high cost of the infrastructure and the costs for the consumers. The final choice was the venerable radio but with the latest digital technology via satellites. As we have seen above, while there was no continentwide satellite radio broadcast in operation in the 1990s, substantial progress was taking place in the underlying technologies, largely through the European efforts for terrestrial digital radio systems. In essence, such a solution adopted for wide-area information dissemination would not only be trail blazing in its sociological mission, it would also have the opportunity to establish a new kind of satellite-based system.

The first major battle was the allocation of frequencies for broadcast digital radio. As we saw earlier, these efforts succeeded in 1992 after some persistent efforts, aided in no small measure by the congruence of the interests of European terrestrial broadcasters on the one hand and those of visionaries like Mr. Samara determined to set up such a system for the developing countries on the other.

The next order of business was of course funding the program. That was no easy task, given the challenge of making any kind of business case for a front-loaded capital-intensive project whose entire *raison d'être* was to improve the lives of listeners with meager purchasing power in the traditional sense of modern market economics. Nevertheless, the project did get financed through private investors, and the industrial activity started in full force in late 1995 for a brand-new satellite radio system. It is a tribute to the strength of Mr. Samara's mission fundamentals and his personal persuasive powers that adequate funding was obtained not just for Africa but in fact for the development of a three-region system, covering practically all of the developing countries of the world in Africa, Asia, and Latin America, with a total population of over 5 billion people and counting!

Several published articles have documented in detail the history, architecture, industrial activities, and actual performance of the WorldSpace system now deployed in Africa and Asia [9, 15–19]. In keeping with the nature of this book, we will only highlight those aspects of this system that are unique and help us understand the achievements and challenges for this system in the context of our business-strategy process.

A large part of the programmatic and industrial activity for this system was led by Alcatel Space at Toulouse, France, in close cooperation with the scientists and program managers from WorldSpace. Other major partners in this endeavor included Matra Marconi Space (now EADS Astrium), Fraunhofer Institut Integriete Schaltungen, Erlangen, Germany, and a still-growing list of premier earth-station and radio-receiver manufacturers all over the world.

System Architecture

It is worth noting that the basic system architecture adopted for the WorldSpace global system in 1995 was not too different from that proposed by Arthur Clarke almost exactly half a century earlier [1]. Just like his concepts, the planned WorldSpace system consisted of three geostationary satellites positioned roughly equally spaced around the world and covering a majority of the land masses below. The three planned locations were 21°E, 105°E, and 95°W. WorldSpace satellites are now in operation from the first two locations over Africa and Asia, respectively, as shown in Figure 10.3.

The WorldSpace system differs in some fundamental aspects from the digital radio systems that preceded it as well as those that followed it (see Table 10.1). While the European efforts at that time were exclusively terrestrial, the WorldSpace system initially was satellite based, with the terrestrial component to follow. The systems that have followed it have targeted automobiles as their primary audience, whereas the WorldSpace system in accordance with its fundamental mission was first aimed at portable use by individuals in their homes or outside. And as we shall see, only the WorldSpace spacecraft design specifically provides the ability to directly receive and integrate content from multiple points in the coverage areas instead of from only a single hub.

The system architecture of the WorldSpace system was heavily influenced by the market it planned to serve. Unlike the relatively homogeneous societies in the West, the continents of Africa and Asia in particular have a large number of countries with significant cultural and climatic differences. Their level of development also varies considerably, a factor relevant in terms of information needs as well as the buying power of the intended users of the system. These factors led to two fundamental decisions in the basic system architecture.

The first decision was to have three coverages in each continent instead of a traditional single coverage of the target market. (As we shall see shortly, this was also partly driven by technical factors.) The primary operating benefit was that the programming contact could now be better tailored to local/regional needs rather than imposing one single package of information over an entire continent with a very diverse combination of cultures, languages, and information needs.

The second factor unique to the targeted markets was the impracticability and high cost of carrying content across several national boundaries to a single uplink station per continent or even within the area covered by each of the three beams. Availability of uplink stations at multiple sites would not only reduce these costs but would also have the flexibility of using content even better matched to the local culture and educational needs.

The size of the total programming package is a compromise between the efficiency of the compression technology, the quality levels desired for different types of programs, and the resultant system impact in terms of spacecraft complexity/size and the available RF bandwidth. In each beam, the WorldSpace system provides a total digital package of 3 Mbps that translates to anywhere from 50 to 100 channels, depending the relative distribution of different types of content.

Each WorldSpace spacecraft payload consists of transparent and processed transponders, as shown in Figure 10.4 [9]. The transparent portion consists of three bent-pipe transponders, one per beam. It functions like a traditional broadcast system, broadcasting one of the two RF carriers with half the programming content that is assembled on the ground at the corresponding hub station. Typically, such hubs can be at a major metropolitan area in each beam or just a single hub station for the entire continent and are convenient points for aggregating content from other broadcasters (e.g., via distribution satellites).

The second half of each spacecraft payload consists of three processed transponders. This part of the payload is designed to receive a large number of individual RF carriers from anywhere in the coverage area, each with as little as a single channel. Within the satellite, these individual channels are extracted from these carriers and combined to form a downlink RF carrier similar to that from the transparent transponders except for one key difference: this RF carrier is modulated with content from many different geographical sites and not from just one hub station. As seen from the receiver, however, both carriers are similar except for the small difference in center frequency and opposite polarizations. The uplink carriers are in the 7,025-MHz and 7,075-MHz band. The downlink frequencies are between 1,467 and 1,492 MHz. Both utilizations are in accordance with the ITU recommendations.

At the time the contract for the WorldSpace satellites was signed in late 1995, the payload and associated spacecraft bus would have been considered truly a high-power one. The output power of each transponder is obtained via two paralleled L-band TWTAs, especially developed for this project. In

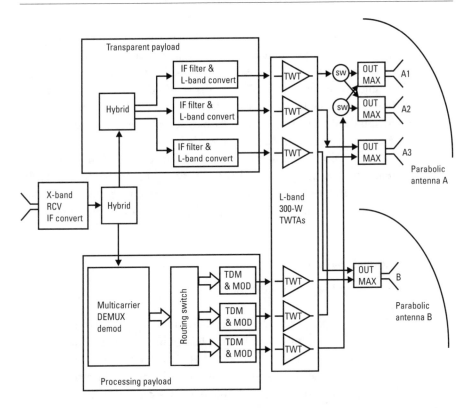

Figure 10.4 WorldSpace satellite payload with transparent and processed transponders. (*From:* [9]. © 2002 ITU. Reprinted with permission.)

principle, more of these TWTAs could have been paralleled, and a single beam with two polarizations could have illuminated the entire continent. However, this approach carried the additional risk of excessive power in output waveguides and would have also forced the same content over the entire area. The use of three beams not only alleviated the technical concerns, it provides additional content flexibility as described above. It also facilitated a common spacecraft design for all three continents.

As we have emphasized throughout this book, the most critical element in a consumer-oriented project is the consumer equipment. In the case of the World-Space system, this is perhaps more true than for any other broadcast system due to the limited purchasing power of the target audience. The development of the WorldSpace receivers benefited considerably from the preceding work carried out for the European system. However, there were important differences. As no satellite-based operational data was available, simulated measurements had to be carried out with the prototype receivers using helicopters to simulate the satellite signal, as summarized in [9]. The finalization of the new compression technology (MPEG 2 layer III), the development of receiver engineering models, and the development of detailed data for chipset developers were all carried out by one integrated team. In view of such little direct relevant experience, two parallel contracts were awarded for the chipsets. A critical decision was to invite reputed consumer electronics companies to build and distribute the final receivers. Before regular production was started, extensive tests were carried out with the newly launched AfriStar satellite. After the initial batches of receivers were produced, efforts have continued to progressively reduce their costs.

WorldSpace has also recently extended the capabilities of the system to a hybrid satellite/terrestrial system. The terrestrial component is compatible with both single and two-satellite configurations. Extensive field trials have been carried out to confirm the system capabilities. Test results are summarized in [9].

Achievements and Challenges

These include the following considerations.

1. The WorldSpace system originated from one man's vision: to make a major difference for fellow human beings by converting "information scarcity to information affluence" [14]. This vision was quite clear and unambiguous, as was its conversion to a mission and in turn to a project based on cutting-edge technology with acceptable risks.

2. The management and implementation of the entire project with significant installations across Africa and Asia was carried out with only modest delays, a tribute to the many industrial partners and to the in-house program-management teams.

3. Apart from being the first satellite digital radio broadcast system, the industrial development also advanced the technology in several areas. Notable examples would be L-band TWTAs, onboard processors for multiple uplinks, and the then-latest compression technology (MPEG 2 layer III). The last item has since been used in well-known MP3 music players. Several parts of the WorldSpace spacecraft payloads, including the TWTAs, also benefited later on one of the U.S. Digital Audio Radio System (DARS) projects.

4. The WorldSpace receivers were perhaps the first-ever major consumer electronics product to be introduced first in developing countries, even though their actual design, initial development, and continued development were carried out by leading companies in Europe and Japan. Accordingly, they did not reap the benefit of either fast demand buildup or quick brand-name recognition, as is generally the case in more advanced societies with much higher buying powers, better distribution, and advertising channels.

5. Until the market builds up to a size large enough to attract a sufficiently large number of paying advertisers, a consumer-based system
has to have other means of recovering its costs and investments. The
more developed satellite and cable television systems in almost all
regions today rely on a combination of subscription services and advertisers. With a few exceptions, none of them are able to provide free-
to-air service, despite many years of growth and market recognition.
The WorldSpace system continues to devote major efforts to reducing
the cost of receivers and is introducing subscription services in selected
areas.

6. The basic architecture of the WorldSpace system, coupled with the
ongoing enhancements for mobile access and continuing receiver
price reductions, allow the realization of the starting vision of alleviat-
ing the information scarcity in certain parts of the world, while
providing a powerful system capability for information and entertainment broadcasts to tens of millions of upwardly mobile potential customers in urban areas within the vast coverage areas of the
WorldSpace satellites.

U.S. Digital Radio Systems

The recent rapid progress in the field of satellite digital radio in the United
States is illustrative of the cliché that has been going around the globe for several
decades now. It goes something like this: while it is the Europeans who make the
inventions, it is the Americans who are often the first to make money from
them. To substantiate this rather sweeping (and unfair) generalization, the most
often quoted examples are the discovery of penicillin and the invention of jet
engines. Obviously, the reality is much more complex, and one can either justify
or repudiate such statements with equally valid arguments on either side. In the
case of digital radio, however, such a statement had some validity, at least until
now. As we have seen earlier in this chapter, Europe not only did lead the way
toward establishing several basic technologies for this field, it also built a major
part of the first such system, WorldSpace, and payloads for the XM spacecraft.
At the same time, so far there is no Europewide digital radio system.

The path to the realization of the U.S. digital radio systems was by no
means smooth either. Several studies were carried out as early in the 1980s, as
we noted earlier [2]. Until early 1990, the focus of almost all such studies was on
the first vision noted earlier in the chapter—the replacement of the unreliable
and expensive HF links. A majority of these studies adopted the familiar single-
channel-per-carrier (SCPC) approach. Coupled with the relatively low power
and mass capabilities of the spacecraft at that time, none of such proposals could
meet the tests of a viable business proposition.

As is often the case, solutions to technological limitations were progressively found; however, the opposition from local broadcasters through their association, the National Association of Broadcasters (NAB), was severe enough to delay by several years any serious consideration of nationwide satellite radio systems. A nationwide satellite-based radio system was painted as depriving the livelihood of several thousand "mom-and-pop" local radio stations. As a result, no FCC licenses were awarded for digital radio systems until almost 5 years after the ITU allocation of the relevant spectra in 1992.

The breakthrough came in late 1996 when sufficient assurances were provided to the so-called small local broadcasters that the satellite-based systems will not compete with them in the local arenas. (The irony of the whole situation, looking backward, is that by 2004, most of the erstwhile small local broadcasters were now part of large nationwide companies, and the so-called local broadcasts were no longer completely local, as some of their programming is created centrally and then distributed after adding some computer-generated "local flavor.")

In early 1997, the FCC announced a spectrum auction for nationwide digital radio. However, this auction was not an open one, it was instead limited to the then-existing six companies, presumably in recognition of their patience awaiting the go-ahead signal for several years. The 50-MHz slice of the S-band spectrum recommended by the ITU, as we noted earlier, was reduced to 25 MHz, to be equally divided between two licensees [20]. The 12.5-MHz spectrum to be given to each of the auction winners could be used for any combination of satellite and terrestrial repeaters use. These decisions were based largely on extensive technical inputs by the parties involved and took into account the prevailing state of the art in satellite and signal-compression technologies.

The two winners of this auction in April 1997 were:

- American Mobile Radio Corporation (AMRC) then in partnership with WorldSpace, now known as XM Radio;
- CD Radio, now known as Sirius Radio.

The FCC also stipulated that both the systems should aim at facilitating common receiving devices for the consumers.

Both systems are now in operation and have some commonalities and differences. Some of these are traceable to early technical and system decisions by the two companies, while the others are due to somewhat different initial strategic and management approaches adopted by the two companies. We will review these pioneering systems in a concurrent fashion.

System Architectures for Sirius and XM Radio Systems

In Chapter 7, we looked at generic block schematics of several major categories of applications of satellite systems. Figure 7.4 provided such a diagram for direct broadcast systems. Figure 10.5 has been derived from Figure 7.4 to provide us with a starting point for looking at the system architectures for the two systems.

Both systems have adopted generally a common approach for the development of the content to be broadcast to the listeners. A good number of the programs are created in-house using digitally recorded and stored music, but with customization provided by specialist disc jockeys (DJs). While such an approach can be quite expensive, it does have the advantage of creating a following for unique programs and genres that should persuade perspective listeners to become long-term subscribers to the system. Both systems also rely to a certain extent on the rebroadcast of other programs, but on a much smaller scale than television broadcasters like DirecTV and Dish networks. Both companies have invested extensively in facilities for live broadcasts; in addition, Sirius has given higher importance to rebroadcast of certain sport channels. Several data services have also been announced by both companies, including up-to-date weather information.

One significant difference until early 2004 was in the area of advertising. While Sirius from the beginning had all channels free of advertisements, XM Radio did have small number of advertisements on some of its channels until

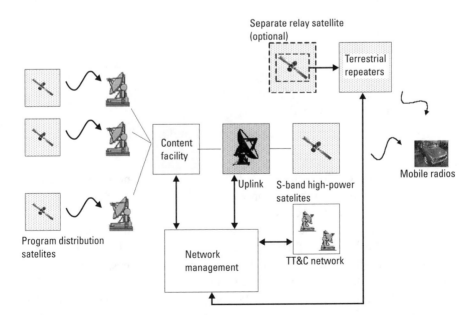

Figure 10.5 Satellite digital audio radio system (SDARS) generic block schematic.

early 2004, when it also moved to all commercial-free broadcasts. Of course all of these practices and policies are essentially market driven and likely will be amended and adapted as more experience is gathered.

In the same vein, several other elements are similar for both of the systems. These include uplink facilities. However, there are major differences in the space segment and in the terrestrial repeaters.

Sirius and XM Space Segments

As noted before, satellite broadcast to mobile receivers is more complex than to static installations. In the latter case (e.g., direct television broadcasts), the receiver antennas are aligned once for all to essentially have an unobstructed line of sight to the satellite(s). Therefore, the only variable element is the additional path loss during adverse weather conditions. While the latter effect is much less pronounced at L- and S-band broadcasts of digital radio, the mobility of the receivers introduces two distinct sources of additional path loss, depending on the position of the mobile listener. The first is the additional losses outside urban areas introduced by variable path blockages due to foliage and other natural and manmade obstructions. The second is the much more severe blockage of satellite signals and multipath effects in major in urban areas. Both of these types of link impairments have a direct bearing on the choice of the system architecture.

Four fundamental measures can be used to counteract such losses in general:

- High-power satellites;
- High-sensitivity receivers;
- Alternative pathways for the desired signal to reach the receiver through some kind of signal diversity;
- Modulation techniques to counteract multipath effects.

While both of the systems have pushed the state of the art with regard to the first two points, they have adopted radically different system philosophies with regard to the latter two.

Sirius has chosen space segment architecture with three satellites in 24-hour inclined orbits around the Earth. While the ability of inclined orbits to provide higher elevation angles, particularly at higher latitudes, has been known for a long time, the Sirius system is the first major operational application of the so-called Tundra orbit with three 24-hour orbit satellites [21]. Figure 10.6 shows the three Sirius satellites following a common track as seen from ground but staggered in time [9, 22].

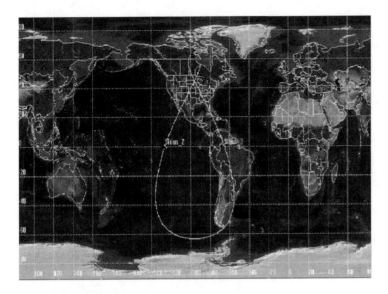

Figure 10.6 Ground tracks for Sirius satellites. (*From:* [9]. © 2002 ITU. Reprinted with permission.)

As explained in detail in the referenced papers [9, 22, 23], the orbital parameters are chosen to make each satellite spend about 16 hours above the equator and 8 hours below the equator. Most of the time, there are at least two spacecraft illuminating the United States, thus providing substantial diversity gains in link budgets. Traditionally, most inclined orbit satellite systems require some kind of tracking for the earth stations accessing such systems. However, the power levels of the Sirius satellites are high enough that the very small antennas in moving vehicles have broad enough beamwidth to access the active satellites without any need for tracking.

The inclined elliptical orbits of Sirius provide a minimum of 60° elevation angle even in the northern parts of the United States, as shown in Figure 10.7 [9, 23]. For comparison, typical elevation angles with geostationary satellites are also shown. These are important and tangible system benefits that translate to fewer blockages and therefore a lesser need for terrestrial repeaters in general. Sirius also opted for the use of a separate Ku-band feed network for the terrestrial repeaters.

The use of inclined elliptical orbit satellites by Sirius requires the TC & R stations to be located outside the United States together with more complicated operating regimes [22]. Figure 10.8 shows a top-level view of the Sirius system architecture with these features [9, 23].

In contrast, XM Radio opted for a somewhat traditional architecture with two geostationary satellites. The driving rationale was one less satellite at the cost of a somewhat larger number of terrestrial repeaters. The two satellites at 85°W

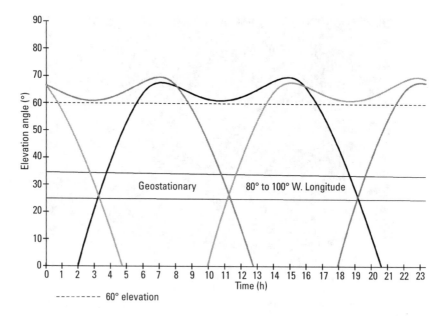

Figure 10.7 Elevation angles at New York for inclined and geostationary orbits. (*From:* [23]. © 2002 American Institute of Aeronautics and Astronautics, Inc. Reprinted with permission.)

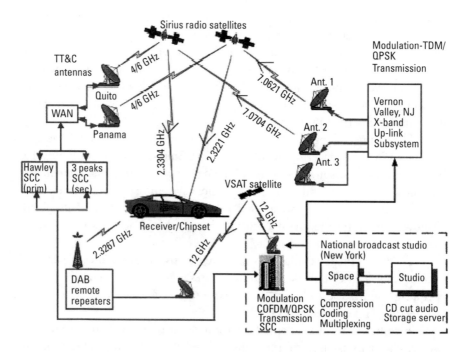

Figure 10.8 Sirius system architecture. (*From:* [9]. © 2002 ITU. Reprinted with permission.)

and 115°W provide full space and time diversity. With the 30° orbital separation, the two satellites provide substantial diversity gains in this configuration as well [24–26]. The frequency plan for the XM system is also different, as it decided to split the programming channels into two signal ensembles with the lower half of the spectrum carrying half of the programming and the upper half carrying the remaining half of the programming, as shown in Figure 10.9. This configuration led to two transponders in each satellite and some resultant link budget improvements.

XM also went a couple of steps further in order to improve availability across the coverage region. First, the EIRP contours were shaped to provide higher power levels at higher elevation angles, where the margin required was higher than that near the equator, thus reducing the overall power requirement. The second step was to further shape the EIRP contours of the two spacecraft in order to enhance the power flux densities in regions with anticipated excessive foliage losses. Figure 10.10 shows the net availability predicted for the total system [26].

Spacecraft Payloads

While the orbital architectures of the two systems are quite unique, the spacecraft payloads are not that different. Both payloads are essentially bent-pipe transparent repeaters creating large levels of S-band powers through paralleled TWTAs. XM uses two transmit antennas. Sirius uses a single antenna fed by four separate quadrants of paralleled TWTAs.

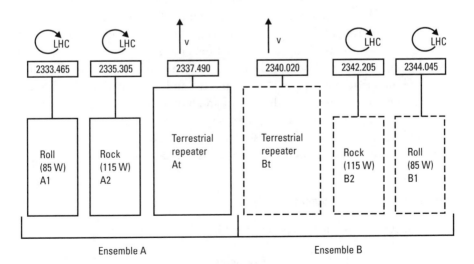

Figure 10.9 XM system frequency plan. (*From:* [26]. © 2004 American Institute of Aeronautics and Astronautics, Inc. Reprinted with permission.)

Baseline

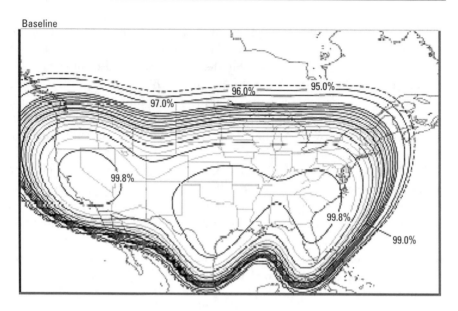

Figure 10.10 XM system service availability. (*From:* [26]. © 2004 American Institute of Aeronautics and Astronautics, Inc. Reprinted with permission.)

SDARS Receivers

Throughout this book we have emphasized the importance of the consumer equipment and total costs. Depending on the region of the world, different aspects assume higher or lower importance. Thus, for the WorldSpace system, the receiver cost was a major factor in the market acceptance and growth of that system. For SDARS systems, serving predominantly cars, the recurring subscription costs have played a greater role than the add-on one-time cost of a receiver to a typical car. Most of the functional units of the receiver are part of the chipset. The major additional items are the antenna/LNA, man-machine interface unique to the product design, and the audio system that is part of the car. The typical receivers are either OEM and look just like any other car radio or can be add-on units that can take a variety of forms. With the ongoing active roll out of both systems, there is a steady stream of excellent consumer devices coming on the market responsive to the diverse needs of the consumer.

Achievements and Challenges

These include the following considerations.

1. The whole process of introducing new technology in the oldest form of broadcasting, namely the classic radio, is on way to become a success in the United States. This success extends across the whole chain,

starting with the regulatory process, albeit delayed, through timely financing, system development, and market deployment of a major consumer product.

2. While both Sirius and XM Radio have demonstrated a very high level of professional management of their respective strategic objectives, XM Radio in the beginning of 2004 had a commanding lead in the marketplace in terms of subscribers. In many ways, the whole process adopted by both of the systems from inception to market launch vindicates several basic tenets of the IBSP presented in this book.

3. Both Sirius and XM Radio can justifiably take pride in their business-oriented approaches in the sense that they resisted the temptation of bragging rights by "reinventing the wheel"; instead, they both harnessed the already developed technologies and adapted several subsystems to their needs, thus reducing program schedule and performance risks as well as lowering the over all costs. Due to the initial leadership by WorldSpace of the XM development phase, this process went further for XM than for Sirius. The outstanding example for Sirius is the successful adaptation of Tundra orbit for its space segment architecture, in the process directly benefiting from earlier efforts [21]. For XM, notable examples of such carry forward of prior efforts include:

   • A common payload developer (Alcatel Space), leading to several WorldSpace developments being directly scaled to the XM Radio needs;

   • Common chipset developers (FhG and ST), which provided tangible benefits through direct adoption of several modules and reduction of the overall development efforts and risks in this area.

4. The experience to date has vindicated the architecture choices and availability objectives adopted by both systems. The ease with which individual terrestrial repeaters can be added, subtracted, or relocated has turned out to be a powerful tool for tweaking the system performance where and when needed.

5. Going forward, there are still some challenges and issues for the two management teams, including:

   • The two systems are still not compatible as seen by the consumer through a common receiver, one of the license requirements. Such a commonality may be realized either through composite chipsets or through software-defined radios (SDRs).

   • Until the systems reach break-even points, both organizations face financial marketplace pressures and almost daily scrutiny by the media, often exaggerating even relatively small hiccups into major management issues. Both managements appear seasoned enough to weather these market vagaries.

- Consumers are being increasingly exposed to all kinds of interactivity everywhere—at home, in their cars, and at work. Current generation of DARS systems are strictly one-way, and the market evolution will need very careful monitoring to pick only those variations and applications that really have the potential to be longer term trends rather than merely passing fads; in other words, concentrate on basic business fundamentals and resist the temptation to go all over the service marketplace.
- Watch carefully for the newly emerging local digital broadcast alternatives and evaluate their impact on the nationwide broadcasts.

MBSat System

The MBSat system, launched in early 2004, is a new and forward-looking entrant in the digital radio field for Japan and Korea. It operates in the upper 2.6-GHz band. So far, only one satellite is planned [27, 28].

This system in many ways could be a harbinger of the evolution of this field from the consumer perspective. A whole range of audio, video, and data services are planned through appropriate consumer devices ranging from pocket-sized hand-held receivers, palm-sized television video receivers (see Figure 10.11), OEM car receivers, and reception capabilities in trains. The terrestrial network is a combination of wide-range and narrow-range gap fillers to suit specific needs in urban areas, shopping malls, and just on the road or waiting on a railway station.

MBSat satellite provides 67-dBW EIRP over 25 MHz of S-band spectrum to run more than 50 channels of audio and video. In addition, the satellite has a 25-MHz Ku-band service link to transmit the broadcast signal to terrestrial repeaters, somewhat similar to the Sirius network. MPEG-4 encoding and advanced audio coding (AAC) compression will deliver CD-quality audio and television-quality video, in addition to various types of data. The system uses code division multiplexing (CDM). The satellite uses a 12-m transmit reflector.

Europe Hybrid System

Earlier in this chapter, we reviewed the pioneering efforts in Europe to make digital radio a reality. It was noted that for a considerable period, there was

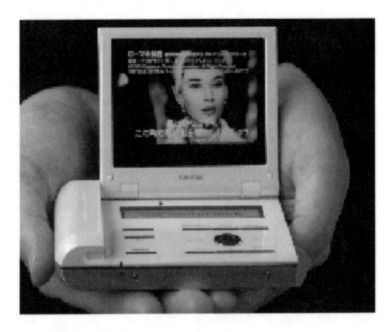

Figure 10.11 MBCO palm receiver. (*From:* [28]. © 2004 American Institute of Aeronautics and Astronautics, Inc. Reprinted with permission.)

resistance to the use of satellites for such services. The reasons were partly technical in the sense that the then-available spacecraft could not economically support the COFDM format adopted as a uniform standard for these services. The other reasons included the fact that local content was considered important and the operating administrations preferred the route of progressively expanding terrestrial-only coverages.

With the demonstrated success of digital radio systems in the United States, there is much greater pressure now for Europe to embark on a hybrid system with satellite(s) covering the continent and terrestrial repeaters augmenting the reception capabilities in urban areas [29].

A couple of years ago, a new company, Global Radio, was formed to embark on such a venture. Unfortunately, in the midst of a weak investment market, this company could not get the necessary financing and recently shut its doors. Nevertheless, it did present possible approaches to meet the somewhat unique requirements in Europe. As an example, multiple beam coverages could respond to the multilingual environment in Europe, while the Tundra orbit [21] could provide higher elevation angles at the even more northern latitudes in Europe than in the United States.

Alternative approaches similar to the XM Radio architecture are also receiving attention, perhaps in conjunction with the current WorldSpace spacecraft in the overlapping orbital arc serving Africa, with some coverage of parts of Europe.

Going forward, a European system will have to address some issues similar to those in the United States and some that are unique. These include:

1. A look at Figure 10.3 on digital radio systems highlights that the available in-orbit spectral resources have to be shared between Africa and Europe. In principle, a total of 40 MHz of spectrum in L-band should be adequate for both the continents, recalling that each similar system over the United States uses only 12.5 MHz. However, any new European system has to take into account the bandwidth already in use by at least the northern beam of the WorldSpace AfriStar and also be compatible with the spectrum already in use in the WorldDab terrestrial system in Europe.

2. While sharing of spectral resources with WorldSpace might favor a geostationary orbit solution, similar to XM Radio, the predominantly northern latitudes of the market would favor Tundra-orbit architecture.

3. Contrary to the market situation in the United States, where there is a strong opposition to satellite-based systems carrying any local content, in Europe there is a certain interest to cater to the local content as well.

4. With the much higher population density in general, the number of terrestrial repeaters needed may be larger than in the U.S. market. This aspect would also favor the Tundra-orbit approach that may require fewer repeaters.

Nonsatellite Alternatives

Finally, it is relevant to mention a couple of terrestrial alternatives vying to take at least a piece of the emerging digital radio market.

We have already seen that a multichannel exclusively terrestrial digital radio approach has not been very successful in Europe due to the prohibitively high cost of covering large countries or continents through single-frequency networks only. However, alternatives means of extending the reach of traditional systems are emerging, and both deserve a brief mention.

Inband over Channel (IBOC) or HD Radio

In the United States alone, there are more than 12,000 AM and FM local radio stations. These can no longer compete either in quality or choices with the new digital radio systems discussed earlier. However, they have certain inherent advantages: low-cost receivers, local content, and broadcasts totally free of any subscription fees.

With the advent of new technologies, there have been numerous efforts to transition these stations to the digital world while still retaining their unique advantages. After years of trials with different technical approaches, in October 2002 the FCC approved the inband over channel (IBOC) standard for such broadcasts.

The basic principle of such broadcasts is shown in Figure 10.12 [20]. In addition to the current AM or FM signals, two lower power digital carriers are inserted within the available spectrum allocations. These digital carriers

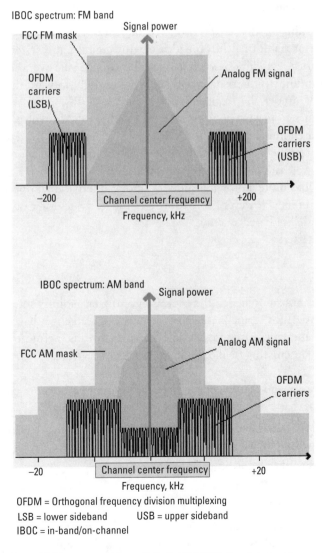

Figure 10.12 Basic IBOC concept. (*From:* [20]. © 2001 IEEE. Reprinted with permission.)

provide means of either simulcasting the same program as the host channel or provide additional music/news or data transmissions. Even though the digital carriers are of relatively lower power, they provide not only better quality but often wider coverage due to their more robust characteristics against interference. Receivers matching the new standard began entering the marketplace in early 2004. It is too early to say how popular these alternatives will be and whether they would give any kind of tangible competition downstream to the satellite-based systems.

Digital Radio Mondiale

In the beginning of this chapter, one of the candidate visions identified was replacement of traditional HF broadcasts with digital radio systems. While in certain situations, satellite-based systems will no doubt take over the related traffic, the community of users below 30 MHz has brought forward their own digital alternative under the Digital Radio Mondiale (DRM) banner [5, 13, 30]. Along the lines somewhat analogous to HD Radio, DRM involves adding digitally modulated information to the current power amplifiers for traditional AM broadcasts. These systems also utilize the same compression technology as satellite radio as well as COFDM in certain situations.

DRM has quite a growing support among international operators as well as manufacturers. However, here also it is too early to forecast any kind of impact on satellite-based systems.

Summary

This chapter has summarized the evolution of digital radio systems, keeping in mind the basic IBSP principles. Achievements and challenges have been identified under individual systems. Some general threads permeating the still-evolving and fascinating field of digital radio systems are:

1. Clarity and consistency of the vision can often be more critical to success than appropriate technology and potential market base.

2. Three unrelated, and still evolving, technologies have enabled the satellite-based digital radio systems a reality. These are:
   - Signal compression;
   - Spacecraft power;
   - Low-cost high-density ASIC chipsets.

3. As is often the case, spectrum availability was the key to the very birth of this field. Regions with no sharing requirements are in a better position to fully exploit the benefits of highest power spacecraft.

4. The early European efforts definitely accelerated the emergence of different enterprises around the world.

5. WorldSpace will perhaps stand alone for a long time as a project brought to fruition almost entirely on the strength of a powerful humanitarian vision.

6. The U.S. systems forcefully demonstrate the power of well-coordinated program implementation and a focused management approach.

7. Future evolution of digital radio systems can be multifaceted. In the developing world, there could be a tremendous growth once the consumer cost barrier is broken through a combination of technology and improving living standards. In the developed word, the whole evolution could become more closely linked to interactivity and automobile telematics.

References

[1]　Clarke, A. C., "Extra Terrestrial Relays," *Wireless World*, October 1945, pp. 305–308.

[2]　Rogers, T., *Space-Based Broadcasting*, Washington, D.C.: National Academy Press, 1985.

[3]　Recommendation ITU-R BO.1130-4, "Systems for Digital Satellite Broadcasting to Vehicular, Portable, and Fixed Receivers in the Band Allocated to BSS (Sound) in the Frequency Range 1400 to 2700 MHz (1994–2001)."

[4]　*Report and Order on DARS in 2310- to 2360-MHz Frequency Band*, FCC97-70, March 3, 1997.

[5]　Deitz, M., and S. Meltzer, "CT-aacPlus, a State-of-the-Art Audio Coding Scheme," *EBU Technical Review*, 2002.

[6]　Faller, C., et al., "Technical Advances in Digital Audio Broadcasting," *Proc. IEEE*, Vol. 90, No. 8, August 2002.

[7]　Thibault, L., et al., "Subjective Evaluation of State-of-the-Art 2 Channel Audio Codecs," *AES Journal*, March 1998.

[8]　Courseille, O., "Satellite Digital Radio Broadcasting: A Today Worldwide Reality; A Challenge for Europe," *19th AIAA International Communication Satellite Systems Conference and Exhibit*, Toulouse, France, April 2001.

[9]　*DSB Handbook*, Radiocommunication Bureau of the ITU, Geneva, Switzerland, 2002.

[10]　Hoeg, W., and T. Lauterbach, (eds.), *Digital Audio Broadcasting*, New York: John Wiley & Sons, 2001.

[11]　http://www.WorldDab.org.

[12]　Shelswell, M. A., "The COFDM Modulation System: The Heart of Digital Audio Broadcasting," *BBC Report*, No. 1996/8, 1996.

[13] Zimmerman, G., "Recent Tests for DRM," *European Digital Radio Conference*, Munich, Germany, April 3–4, 2003.

[14] Samara, N., *Speech at NASA's 15th Annual Conference on Continual Improvement and Reinvention*, Alexandra, VA, April 27, 2000.

[15] Campanella, S. J., "Asiastar: A Digital Direct Broadcast Satellite," *Pacific Telecommunication Review*, 3rd Quarter, 1998.

[16] Chandrasekhar, M. G., "Digital Radio: Promises and Pitfalls," SSPI Mid-Atlantic Chapter presentation, July 29, 2003.

[17] Courseille, O., P. Fournie, and J. F. Gambart, "On-Air with the WorldSpace Satellite System," *48th IAF Conference*, Turino, Italy, October 1997.

[18] Courseille, O., and P. Fournie, "WorldSpace: The World's First DAB Satellite Service," *Alcatel Telecommunication Review*, 2nd Quarter, 1997, p. 102.

[19] Sachdev, D. K., "Digital Sound Broadcasting Goes Global," *International Radio Symposium*, Montreux, Switzerland, June 10–13, 1998.

[20] Layer, D., "Digital Radio Takes to the Road," *IEEE Spectrum,* July 2001.

[21] Ashton, C. J., "Archimedes-Land Mobile Communications for Highly Inclined Satellite Orbits," *IEE Conference on Satellite Mobile Communications*, Brighton, England, September 1988, pp. 133–137.

[22] Briskman, R., and R. Prevaux, "S-DARDS Broadcast from Inclined Elliptical Orbits," *52nd International Astronautical Congress*, Toulouse, France, October 1–5, 2001.

[23] Briskman, R., "DARS Satellite Constellation Performance," *AIAA Satellite Systems Conference*, Montreal, Canada, paper 2002-1967, May 2002.

[24] Michalski, R., "Method of Calculating Link Margin in a Satellite System Employing Signal Diversity," *AIAA Satellite Systems Conference*, Montreal, Canada, paper 2002-1856, May 2002.

[25] Michalski, R., "An Overview of the XM Satellite Radio System," *AIAA Satellite Systems Conference*, Montreal, Canada, paper 2002-1844, May 2002.

[26] Snyder, J., and S. Patsiokas, "XM Satellite Radio—Satellite Technology Meets a Real Market," *22nd AIAA International Communications Satellite Systems Conference & Exhibit 2004*, Monterey, CA, paper 2004-3227, 2004.

[27] "MBCO: A True Ubiquitous Broadcasting System," *APSCC Newsletter,* Winter 2004.

[28] Yamaguchi, Y., "S-Band Digital Multimedia Satellite Broadcasting Services for Personal & Mobile Users in Japan," *22nd AIAA International Communications Satellite Systems Conference & Exhibit 2004*, Monterey, CA, paper 2004-3213, 2004.

[29] Kozamernik, F., N. Laflin, and T. O'Leary, "Satellite DBS Systems," *EBU Technical Review,* January 2002.

[30] Stott, J., "DRM: Key Technical Features," *EBU Technical Review*, March 2001.

11

Future Evolution

> It is not the strongest of the species that survive, nor the most intelligent; it
> is the one that is most adaptable for change.
>
> —Charles Darwin

It is perhaps a worn out cliché by now to start a chapter on future evolution with
a quote from Charles Darwin. In modern times, technologies come and go at an
ever-increasing rate, and their active reign is not always long enough to talk
about evolution per se. However, when it comes to the satellite industry and its
global community, there is reasonable justification to do so. This technology has
been around for nearly half a century and, despite some recent setbacks and a
few self-inflicted wounds, there is sufficient reason to believe that, even in the
twenty-first century, we will continue to rely on this medium in some fashion or
other.

In preceding chapters, we addressed different stages involved in taking a
business strategy for a satellite system to fruition and the appropriate approaches
to make each such major link in the chain successful and responsive. In this con-
cluding chapter, we will attempt a top-level prognosis for the future.

Intrinsic Strengths and Adaptability to Change

The longevity and relevance of a technology or even an industry as a whole
depends largely on a complex combination of its intrinsic strengths and its
adaptability to changing times and consumer needs. In the real world, the
strengths may not be fully available in the beginning but can become practical
with advances in associated techniques. The value of the intrinsic strengths can

also change with the dynamism in the external competitive environment. Along the same lines, the flexibility and adaptability to change on the part of the industry leaders and managers can also influence its growth. An adverse outcome in both of these fundamental areas often leads to a steep decline unless the whole or parts of the industry are propped up for some time by artificial means.

Intrinsic Strengths of Satellite Systems

The principal intrinsic strength of satellite systems is their ubiquity. It is as fundamental and unique to the Earth and its inhabitants as the Sun and the Moon are. Unlike any other medium, whether wired or wireless, the lofty heights from which satellites operate give them unparalleled bilateral *radio* visibility across large areas of the Earth. However, for a fairly long period in the beginning, this unique strength could not be fully harnessed due to limitations in technology. As an example, when the first satellites were lofted into space in the early 1960s, the ubiquity of this exciting new medium was only *in principle,* as the satellite radiated powers were so low that they could only be made use of with really giant antennas of the type shown in Figure 11.1. It took several decades of technological advances, system development, and international agreements before this strength of universal ubiquity could be translated to DTU services in homes and automobiles almost three decades later [1, 2]. Nevertheless, in the intervening years, the intrinsic ubiquity of satellites was harnessed in other ways to meet a variety of market needs. A few examples follow:

Figure 11.1 Andover horn antenna.

- Even with the early generation satellites, a number of previously inaccessible or less developed areas of the world were connected to the rest of the world through high-quality telecommunication links, albeit still with fairly large antennas. The landmark achievement here was that such links were established without necessarily awaiting the traditional outward expansion of intervening networks.

- As the number of earth stations grew all around the world and the satellite capabilities improved, the ubiquity of geostationary satellites provided unique switchboards in the sky, literally bringing together hundreds of telecommunication and broadcasting networks across oceans.

- The global positioning system (GPS) started as a military system but today is by far the most ubiquitous application of the satellite medium, providing precise location coordinates anywhere on land, sea, or air, and in the process creating a $10 billion industry for the user devices alone.

- From the very beginning, the ubiquity of both GEO and lower orbit satellite systems had considerable strategic importance for defense systems. At several stages of the industry evolution, such applications have in fact accelerated the technology-development process with substantial payoffs for the commercial applications as well.

Adaptability to Change in the First 50 Years

In its first 50 years of existence, how adaptable and receptive has this medium been to change? Did the changes happen only in response to external stimuli or were they also internally generated by individual leaders or the community at large in anticipation of the future needs? Have the changes always been successful and beneficial to the consumer? All of these questions, though important and relevant, are difficult to answer for an industry made up of literally hundreds of entities spread across the world and over several decades in time.

Overall, the satellite community has shown an impressive combination of innovation, grit, and responsiveness. These attributes cover all aspects, ranging from technology, services, and organizational structures to business strategy. It would also be fair to say that there have been instances where changes have been too slow in coming or led to severe setbacks. We substantiate these assertions with a few examples over the long half century of evolution.

- If past precedents for new developments had been allowed to prevail, the landmark technology demonstrations with the very first satellites in the early 1960s would have been followed only by a gradual introduction of such systems, perhaps on a bilateral basis between the more advanced nations. Substantial credit belongs to the then–technology leaders, regulators, and governments that saw a unique global unifying role for this technology and painstakingly evolved a structure for a truly international operating organization, INTELSAT [1]. It not only assured equitable and uniform access on a true global basis, but also put in place a sound framework for a steady infusion of international capital for the technology development as well as for the space and ground infrastructure deployments worldwide. This unique structure for an international system served its purpose remarkably well for several decades and was emulated elsewhere for several other similar needs as well.

- It is ironic what passage of time can do. What was in the beginning an example of positive change was within a few decades seen as a restraint and a liability. As the satellite systems expanded to the national and regional arenas, the business leaders outside the community of mostly governmental owners of the international consortia such as INTELSAT began to perceive them as barrier to true entrepreneurship. The *incumbent* group of owners was often painted as an exclusive international club of government entities that was slow to respond to the end users and did not really open up the market to private operators. Eventually, albeit slowly, yet another change did happen, though it was pushed from the outside by leaders like Rene Anselmo, as highlighted in Chapter 4. This transformation from treaty organizations to traditional company structures progressed slowly, and by 2004 almost the entire industry had been privatized. However, it is too early to make any judgment that the original highly successful international structures had really lived out their usefulness and reverting back to the traditional business modes at par with other fields was indeed a change in the interest of the consumers.

- As technology advanced, true ubiquity in terms of direct access to individual mobile users became an attractive challenge, particularly to those segments of the industry unconstrained by the structures and policies of large telecommunication network operators. The fast pace at which dramatically new concepts came out to meet such objectives was a direct tribute to the spirit of innovation and the intense desire for dramatic changes bordering on paradigm shifts. As we saw earlier, some major satellite mobile systems did get built, but unfortunately fell short of their objectives due to a complex combination of their business strategies and the fast-paced changes in external environments. Nevertheless, the lessons learned by the community from these commendable motivations for change will no doubt make future such projects more successful.

- Perhaps the field of direct broadcasting best illustrates how the industry has successfully made the changes necessary to respond to the customers' needs. After some costly missteps, compounded by relatively immature technologies, direct broadcast of television programs became successful through some innovative, outside-the-box thinking by a little company in Luxembourg in the 1980s. The next major step forward was through a powerful symbiosis of spacecraft and compression technologies by the all-digital broadcasters in the United States. Equally pioneering thinking has been responsible for the evolution of digital radio broadcasting, as summarized in Chapter 10.

Evolution in the Coming Decades

With this backdrop, we will now attempt a prognosis on how the satellite systems and the industry could evolve during the next couple of decades or so.

The three pillars of the future evolution of satellite systems are technologies, services, and business strategies. As stated in a somewhat different context in Chapter 1, technologies are not end objectives by themselves but are indeed critical enablers of competitive services. And, finally, success in the realization and operation of such services depends substantially on the underlying business strategies and on management leadership at all stages of the evolution.

Technology

At the top level, technology for satellite systems has two broad measures or metrics for advancement—launch vehicles and spacecraft bus or platform in one group and overall system or network and spacecraft payload in the other.

The first group represents a general industrial capability and is not directly linked with specific systems. Launch vehicle capability provides an upper limit for the total mass of the spacecraft and its physical size. This generally translates to an upper bound on the space segment capabilities in terms of total power, channel capacity, and the largest antennas. Currently, three large launch service providers from Europe and the United States provide lift-off capabilities approaching 10 tons in certain cases, considered adequate for the foreseeable future. Close behind these large providers are at least three mid-level providers, led by China and followed by Japan and India. In addition, there is substantial activity at the low end, with a lot of innovation but with considerable churn in the longevity of such enterprises. Several launch vehicles of all sizes continue to benefit from the missile development and other government programs.

The second important element of the first group is the spacecraft bus. Figure 7.11 identified the key building blocks and their functionalities, complexity, and impact. Two of these have a more direct impact on future evolution. These are the power and propulsion subsystems. The former controls the total spacecraft capability, while the latter controls its useful lifetime. Solar cells continue to advance to the point that often the limiting factor in spacecraft design is the thermal subsystem's ability to radiate out the dissipated heat away from the body of the spacecraft. A combination of innovative radiators and higher efficiency payloads will no doubt remove these barriers to still higher power levels.

Propulsion systems have been trying for several decades to break away from chemical technologies and move on to substantially more efficient electric technologies. There is progress here, but it is relatively slow, especially in terms of in-orbit reliability. As for many other subsystems, the reliability challenge for such subsystems will also be met, thereby finally realizing the substantial economies of electric propulsion through reduced launch mass and hence costs.

At the system and payload levels, technology has indeed been an important driver from the very beginning. Maral and Bousquet, authors of several excellent books, have captured this synergy between technology and services over several decades in a remarkably eloquent graph reproduced in Figure 11.2 [3]. It gives

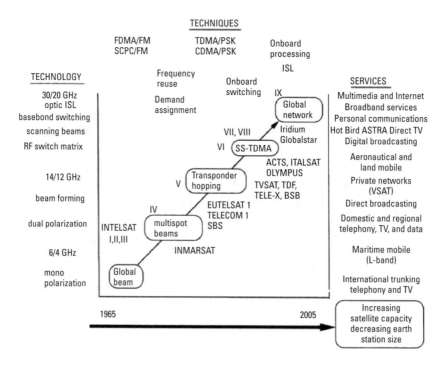

Figure 11.2 Technologies as prime movers of progress 1965–2005. (*From:* [3]. © 2002 John Wiley & Sons, Inc. Reprinted with permission.)

center stage to the techniques, with the key technologies on the left and the driving or beneficiary service groups on the right. A pair of important metrics that increase satellite capacity and decrease earth-station size are shown along the x-axis as a function of time. What is somewhat implicit in this is that the normalized unit cost of service has also decreased consistently along the same time frame. At the heart of these achievements are at least two submetrics: lower cost ground segment through higher power from the satellites and lower cost per transponder through larger satellites. We should also recognize in the same breath the contributions of compression and ASIC technologies through greater system throughput and lower cost and higher reliability ground equipment for consumers.

Looking into the future, there is a strong school of thought that believes that bigger and more powerful satellites, despite their risks of high upfront investment, are still the way towards sustained competitiveness into the future. Large spacecraft or space platforms, as they were first called, are by no means a new concept [4]. These were envisioned as large three-dimensional structures housing a large number of antennas, a condominium in space if you like. In light of the tremendous challenge of optical fibers starting in 1980s, however, such concepts went out of favor in the commercial world with emphasis turning to buying off-the-shelf spacecraft instead, although large military spacecraft continue to be designed and built.

There is now renewed interest in spacecraft larger than the current average from several perspectives. Thus, Takashi Iida and his colleagues [5] approach this subject primarily from the perspective of compatibility with extremely large capabilities of global optical networks. They reason that, paradoxical as it may seem, the tremendous threat from optical fibers indicates that satellite systems have to be larger than they are now, just to capture 1% of the backbone network capacity. The objective of their detailed work is not to immediately launch operational projects with large spacecraft but to start the needed technology programs well ahead of time. Figure 11.3 from their referenced paper captures the general direction in which development work should proceed through several generations. The two parameters on this chart are in-orbit mass and satellite capacity. They highlight that the two major broadband programs that were scheduled to be launched in 2004, iPSTAR and Spaceway, break away from the slower gradient of previous generations, and this trend should be exploited for still larger capacity spacecraft with relatively less corresponding increases in in-orbit mass. Using terminology somewhat akin to cellular generations, they envision the 2G series of spacecraft to have throughput of as high as 500 Gbps.

A second and complementary perspective can be obtained from another contemporary paper by Jean-Didier Gayrard [6]. Approaching the payload evolution from the ongoing transformation in the dominant service groups, he argues that the satellite systems will have to contend with a much higher level of

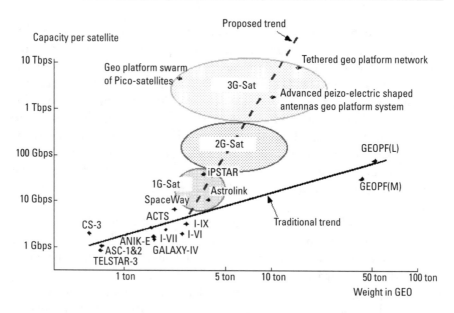

Figure 11.3   Possible evolution of spacecraft size and capacity. (*From:* [5]. © 2002 American Institute of Aeronautics and Astronautics, Inc. Reprinted with permission.)

complexity as they transition from the C-/Ku-band systems for telecommunication and television to those designed for multimedia and Internet access.

This complexity emanates from two factors. The first is the substantial increase in the number of accesses per satellite, and the second is the decrease in the information quantum or elemental bandwidth. Both of these trends translate to systems with a much higher number of receive and transmit beams and the need to carry out real-time switching at lower and lower levels of information hierarchies. It is worth noting that the concomitant need to create small beams mandates a single large spacecraft with large reflectors, rather than the colocated clusters currently in use, to increase the availability of the total number of transponders at one location. Just like Iida, he envisions three generations of satellites, but driven by the switching levels in the payload, possibly reaching the kind of capacity levels per satellite projected by Takashi Iida from macro-level considerations. After a fairly detailed and logical treatment, he names the winner as "a high-power GEO satellite working in Ka-band with regional multi-beam coverage and an onboard processor" [6]. He envisions three generations to evolve around this general concept, shown in Figure 11.4, with each generation having higher capacity than the previous one and reaching the user through smaller devices. The "final" destination is to have one beam per user with onboard processing (OBP) and regenerative technologies. It is interesting to note that Mark Dankberg of Viasat recently made a similar observation

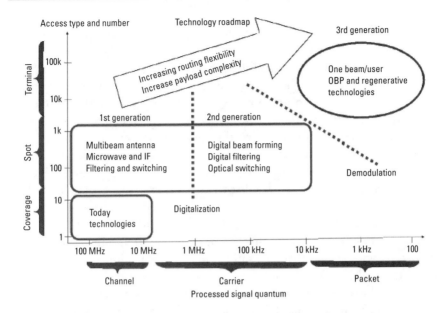

Figure 11.4 Possible evolution of payload generations. (*From:* [6]. © 2002 American Institute of Aeronautics and Astronautics, Inc. Reprinted with permission.)

that "with an appropriate customer-centric system approach, we can have within five years satellite capacities as high as 100 Gbit/s with Ku/Ka bands alone, with terabits of capacity in a region. Such a capability will really make broadband widely available and at a really affordable cost" [7].

While on the topic of Internet access and potentially large-capacity satellites, it is also relevant to recognize the ongoing developments in the military sector, under a variety of names and acronyms, the most well-known of that nowadays being transformational communication architecture (TCA). Recognizing the importance of getting the right information to the right person when needed, the vision of TCA is to create "Internet-like transport architecture between space, air and ground nodes" [8]. Such an architecture will seamlessly interact with systems of different generations, some deployed and others on the way, finally creating a truly global access through high-capacity links in space utilizing laser and RF communications. Operating as a global information grid (GIG), it also envisions fast-accessible storage facilities where needed. Those who have been in the commercial part of the industry long enough will recognize in this major initiative by the U.S. military potential fruition of several decades of far-out ideas that never could go beyond concepts or technology development. Thus, intersatellite links were active topics even in the 1980s, and some of the very first in-orbit tests and concepts were reported as early as 1978 [9, 10]. At a minimum, the TCA-related activities are worth watching, as many of the concepts may have commonality with the commercial sector in the not-too-distant future. The hardest part here is perhaps keeping track of all of the acronyms!

In the context of a potential justification for large-capacity spacecraft, we should not ignore that we have a limited spectrum, particularly if every operating system wishes to be in the most favored bands for high capacities, the Ku- and Ka-bands. Furthermore, the move toward smaller user antennas mandates moving the spacecraft further apart, thus reducing the overall capacity of the geostationary orbit. In this regard, it is worth looking at other orbits as well, particularly the elliptical inclined orbits. One effort, specifically targeted at increasing the total orbital capacity, has been led for many years by Draim and his associates [11, 12]. Starting with the classical inclined orbit systems, like Molniya and Tundra, this group has evolved a series of new concepts aimed at maximizing the number of satellites that can be deployed around the Earth without interfering with each other. Adopting, for a change, interesting acronyms and names like COBRA and Tear-Drop, they essentially extend the 2° spacing around the geostationary orbit on several arcs around the Earth (see Figure 11.5). Being a former Navy aviator, Draim envisions thousands of satellites marching in formation around the Earth, multiplying the capacity capabilities several times over. While such formations may apply only for ground networks with larger tracking antennas, the principles enunciated and tested certainly deserve to be considered when needed.

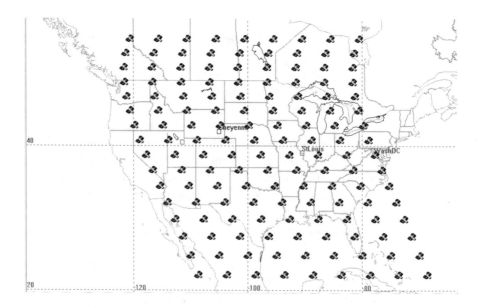

Figure 11.5 Clusters of satellites with inclined elliptical orbits. (*From:* [12]. © 2002 American Astronautical Society. Reprinted with permission.)

For a book on business strategies, we have perhaps waxed eloquently enough about new technologies and related system concepts, and it is appropriate to turn toward ways of deciding which technology to use and why. As a first step, we once again turn to de Weck and his associates at the Massachusetts Institute of Technology and at the Communication Research Laboratory in Japan. They have recently extended their system optimization approach, discussed in Chapter 6, to provide a quantitative tool that can enable selection among competing technologies [13]. Such selections can be made on the basis of criteria like overall life-cycle costs and capacities. It is interesting to note that in the published paper, the group evaluates four emerging technologies: optical intersatellite links, asynchronous transfer mode (ATM), large deployable reflectors, and digital beam forming as applicable to LEO systems. However, the methodology and the associated simulation are generic enough to have wider applicability. Like all otherwise powerful models, its results are as accurate as the inputs with regard to the costs and system benefits associated with each aspiring technology.

Services

We now address the second pillar for future evolution—the types of services in which satellite systems can be competitive in the future. While doing so, we have to keep in mind the ultimate consumer, who will more and more directly interface with the satellite systems and to a decreasing extent through the traditional telecommunication networks. The consumer of tomorrow is likely to be increasingly mobile, preferring a laptop, mobile phone, and personal digital assistant, rather than being tied to fixed-line phones and desktops. Devices combining the functionalities of all such devices, often including video and imaging capabilities, may soon become the primary interfaces almost everywhere. Direct Internet protocol (IP) interface, or *Internet access*, is already becoming a necessity. Transportation systems in the future will no longer be just for traveling from a point A to B, they will be also expected to enable consumers to keep constantly in touch, enjoy real-time entertainment, or get work done while on the move. The cars of tomorrow will no longer have just phones or radios, but total integrated systems plugged into telecommunication, safety, entertainment systems, and broadband highways.

This ongoing transformation of services is indeed mind boggling for any planner or strategist. Fortunately, a good proportion of the services of the future can indeed be provided by the satellite medium, either directly or indirectly. We will take a brief overview of these changes and opportunities using the alternative infrastructures evaluated in Chapter 7 in the context of identifying common engineering building blocks. Wherever applicable, we will draw attention to transponder utilization and forecasts, shown in Figure 11.6, from a recent

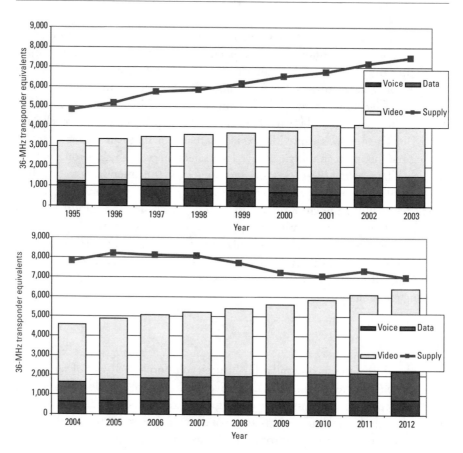

Figure 11.6 Trends and forecasts for transponders. (*Source:* Futron Corporation, 2004.)

presentation [14]. This figure is an extension of Figure 1.3(d) and now includes forecasts as well.

Telecommunication Services

International and domestic trunk services of the types shown in Figure 7.2 were the prime engines for the early evolution of telecommunication services via satellites through a capability superior to that offered by the then-copper cables across the seas and large land masses. The higher capacity of optical fibers first challenged the trunking routes across land masses and a bit later across the seas. This competitive challenge was more successful in land-based thick routes. For long international routes and connectivity networks of the type shown in Figure 7.2(b), the satellite systems withstood their ground to a certain extent, especially to relatively interior destinations and new links that continue to emerge in new territories. More recently, the point-to-point symmetric voice traffic has been outpaced in growth rates by the asymmetric IP links in many parts of the world. Nevertheless, despite a phenomenal increase in the overall telecommunication pie, the relative share as well as the absolute number of transponders for voice has declined, as can be seen from the top graph in Figure 11.6. However, the combination of voice and data is forecast to hold on to a steady and modestly rising share through the coming decade. Such global numbers, however, tend to mask a very impressive growth in VSAT networks [see Figure 7.2(a)], which generally utilize modest space segment resources but can have relatively large terrestrial networks for a variety of needs.

In terms of technologies highlighted earlier, this sector is not the primary driver. However, greater agility in terms of reconfiguration of capacity and coverages to match peaks in demand (e.g., for government needs) can indeed benefit this sector and could even counterbalance the erosion of telecommunication market share by other media.

Broadcast Systems

Direct broadcast systems for both television and radio are expected to grow at an impressive rate across all of the components captured in Figures 7.3 and 7.4. Both of these services continue to push the space segment capabilities toward true and practical ubiquity. Coupled with equally impressive progress in compression and ASIC technologies, this segment can be expected to continue to enhance its service offerings as well as to expand across the globe with hopefully lower total consumer costs. The television segment was the first to reach a critical mass, while the radio sector is still relatively new. Television broadcast at the Ku-band to fixed-home installation has greater external environment pressures through competition from cable systems with steadily improving capabilities and expanding reach. By contrast, radio broadcasts to the vast and growing populations of cars and other vehicles across the globe has less competitive pressures from other media. Both types of such services are likely to see various forms of integration with other services in their respective domains. Figure 11.6 shows the anticipated growth for all kinds of video services. It is to be noted that the traditional measures of growth via number of transponders tend to break down for broadcasting systems, particularly for radio.

Broadcast systems represent perhaps the strongest offering of the satellite medium and not surprising the most successful today. Even a single satellite system gets high market evaluations, depending on the number of subscribers and their average payments. Innovative ways continue to be evaluated to make these services even more attractive. Many of these involve integration or *bundling* with other services for possible synergy, as referred to in earlier chapters. The evolution of digital radio systems toward selective narrowband video was covered in Chapter 10. Some interesting projects are considering enabling next generation cellular phone systems to receive such video broadcasts on their cell phones, thus possibly creating a new business strategy [15].

In terms of technology, broadcast systems have played a major role in pushing the spacecraft power-generation capabilities, and this trend can be expected to continue, thus enhancing the ability to receive broadcast signals even under heavy obstructions. However, the greatest benefit in terms of worldwide expansion of this flagship service will accrue through innovative system and technology solutions that reduce the total cost to the consumer.

Broadband Systems

It is interesting to note that as recently as 2001, the consensus of analysts and conference panels was that satellite systems can capture only a small portion of the market not within the reach of either DSL or cable modems. In other words, the satellites were being relegated to make do with the remnants of arguably one of the hottest markets of the future. In contrast to this, a very recent report by Frost & Sullivan states that satellite broadband can be competitive [16].

What changed in such a short span? To a large extent, the history of television in the late 1980s is possibly getting repeated. As noted in Chapter 2, in the early days of satellite broadcasting, such systems were expected to pick only the rural audience away from traditional broadcasters. However, only when a combination of sound business strategies and technologies permitted the satellite medium to complete head on with cable systems in the heart of urban markets did satellite television broadcasting come into its own to become a full-fledged competitor.

For broadband, a somewhat similar path is being followed, although the overall cycle time may be shorter, like so many other fields in this century. The initial broadband offerings via satellites were extensions of Ku-band broadcast systems and were not considered attractive due to a variety of reasons, principally the high consumer equipment cost and subscription services. That accounted for the pessimistic forecasts until quite recently. Today, with large multibeam high-capacity Ka-band systems like iPSTAR, SpaceWay, and Wildblue about to be launched, there is a real prospect of truly competitive offerings in this segment. The technology developments summarized earlier also support this likely trend. We will address the underlying risks in the next section.

Mobile Systems

What was said about the broadband trends perhaps applies with a greater force to mobile services via satellites, with some notable differences. As we have discussed at several places earlier, the satellite systems did make bold and innovative attempts in the 1990s to compete head on in the heart of the mobile marketplace but did not succeed. After some regrouping and technological advances, the industry is now attempting to get back into the mainstream business via two broad avenues.

The first is the steady progress in global mobile networks, led by INMARSAT. This network will soon offer through its fourth generation satellites, boasting mobile broadband capabilities with speeds up to 432 Kbps. Admittedly, due to the size of the consumer equipment and possibly tariffs, this is not quite in the heartland of the mobile market led by generations of cellular phone systems, but it is a demonstration of the intrinsic strength of ubiquity simultaneously with an impressive transmission speed. Whether it is classified as broadband or mobile, the marketplace will decide.

The second avenue is bolder, as it does directly target the ubiquitous cell phone user (see Figure 7.6). Exploiting some of the very same technology as broadband but at lower frequency bands, such systems aim to provide satellite access through the same handset as those used for accessing a cellular tower a few kilometers away. This is becoming possible through a combination of technology and regulatory advances. Technology provides sufficiently large powers feeding large reflectors, creating a cellular pattern of their own on the ground. Figure 11.7 shows such cell patterns together with the cellular patterns of the terrestrial cellular networks wherever available [17]. The regulatory advance allows the satellite-allocated subbands not in use in a particular cell to be dynamically

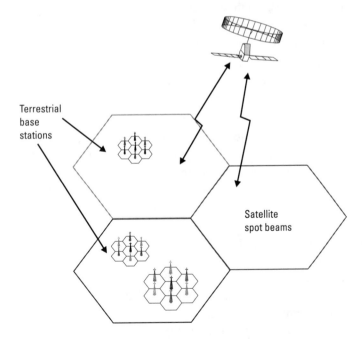

Figure 11.7 New integrated mobile systems. (*From:* [17]. © 2003 Mobile Satellite Ventures, LP. Reprinted with permission.)

allotted to the terrestrial network. This real-time symbiosis between satellite and terrestrial networks promises a truly ubiquitous service. The concepts as well as marketing challenges are significant here, but they appear to be well in hand and several commercial and government ventures are planned in this arena.

Summary of Services and Technologies

With these quick overviews for technologies and services, we are in a position to summarize the key points as inputs to the final section on business strategies.

1. Video distribution of different types will continue to grow and dominate over telecommunication links in the fixed satellite service (FSS) sector. Better C- and Ku-band bent-pipe spacecraft with declining cost per transponder will continue to evolve.

2. Both video and radio broadcasting systems will continue to grow in current markets and in new markets. L- and S-band radio broadcast has two intrinsic advantages: users can be fully mobile and the installed base of radios is larger than the number of homes in most markets. On the other hand, Ku-band for video broadcasts has much larger bandwidth for multimedia than L-/S-bands for radio.

3. The new GEO-mobile systems could be the second major attempt to get this service segment right and competitive to the continually advancing capabilities of the cellular networks.

4. Broadband at the Ka-band is entering a critical test phase with the Wildblue, Spaceway, and iPSTAR systems. The general trend is to make such services attractive via spacecraft with as high capacity as possible and user terminals smaller and more affordable.

5. An interesting combination of items 3 and 4 suggests a future where multimedia services will be provided to mobile users via L-/S-band spacecraft while wider band multimedia and other broadband services to homes and businesses will be provided at the Ka-band.

Business Strategies into the Future

We now consider the final of the three factors for future evolution: business strategies. Unlike technologies and services, we have an interesting paradox here in the sense that some things need not change while others must.

Going forward into the future, the fundamentals of business strategy development as discussed in some depth in Chapter 4 do not change. Any new venture or a major project still needs a clearly defined mission that recognizes the stakeholders, principally the customers and the investors. The external environment analysis is likely to be even more important, given the increasing pace of the changing landscape of competitive alternatives. Given the increasing range of alternative business, system, and technical alternatives, the process of developing strategic choices and winnowing them to the one most responsive to the mission will remain the most critical activity and can determine several years in advance the probability of ultimate success. The only way to get it right is to have a process that is as informed, as in depth, and as interactive as possible. The IBSP developed in some detail in this book is one proven way of ensuring that all of the knowledge relevant to this process is brought to bear in an orderly and timely fashion at every stage.

What is also not changing is the importance of a sound business plan with clear objectives. Depending on the audience, this plan should be representative of the organization as it is or as it aspires to be. And, finally, a synergistic system planning and engineering process is even more critical than before to build an efficient infrastructure responsive to the marketplace.

However, there *is* also a need for some fundamental changes. These changes in many situations will test the depth of leadership and managing acumen, as discussed in Chapter 9.

The most fundamental change is that several layers between the top management and the ultimate customer are fast disappearing. The days when the satellite systems were carriers' carriers, often many times over, are fast dwindling. Therefore, the simplistic yardstick of success in terms of the lowest cost of a transponder is no longer the most critical one. Rather, more and more services are going to be judged by the ultimate consumer's satisfaction, with a mix of criteria as seen from his or her perspective alone. The consumers are going to see more and more choices and can therefore collectively determine the ultimate success of a system in the marketplace. This requires that the top leaders and managers are not only fully aware of what the consumer needs but are also willing to act accordingly. As illustrated through a couple of simple scenarios in Chapter 7, it is no longer appropriate to first design a generic satellite and then let the cost of the transponder capacity determine what the users will have to spend on their equipment and subscription charges. Instead, the system has to be built from the consumer upwards, and the satellite should be the last, and not the first, item in the chain to be frozen in its configuration. In other words, as was emphasized earlier in this book, the top managements have to make a conscious move from being satellite centric to consumer centric.

The second major change in many ways is implicit in the first change but is important enough to be highlighted in its own right. This is the fundamental shift underway in the service mix for the satellite systems in the future. We have already seen that through a recent analysis of transponder utilization captured in Figure 11.6.

As discussed earlier in this chapter, this shift is happening due to a combination of almost overpowering competition from optical fibers in traditional telecommunication links and the ability of the improving satellite technology to finally implement its intrinsic strength of ubiquity with impressive agility and mobility. Once again, such advances in industry have to be matched with the appropriate mix of leadership and management talents.

The third major change is often necessary whenever there is a paradigm shift in the industry. Specifically, when we move from the relatively safe haven of long-term leases for transponders to the constantly changing churn in the number of subscribers, we inevitably increase business risks of a variety of types. In several chapters, we have seen the impact of such risks, sometimes due to misjudging the market needs and on other occasions missing the windows of opportunity. However, the future may bring additional risks that need to be recognized.

The two major new areas highlighted earlier were broadband and the new integrated mobile systems. Both envision reducing costs per unit of service by maximizing the total capacity of spacecraft. This is proposed to be achieved through large spacecraft with impressive technologies, each of which is more efficient yet increases the nonfungibility of the space segment. Therefore, any misjudgments in market share availability can have serious consequences. In other words, a question needs to be asked: while recognizing what the customer needs, are we once again becoming satellite centric?

The proper way to handle such risks is not necessarily to shy away from promising technologies, but to have an organizationwide risk-reduction program from the very start. This should be the domain of all involved in a process like the IBSP, not just the engineers or marketing executives. And if at any stage the answer is not to invest because the risks are too high, the proper management response should be not to shoot the messenger but instead to devote greater energy to understanding and, if feasible, retiring the risks.

Summary

In summary, for this chapter as well as for this book itself, it is fair to conclude that the satellite industry has indeed met Charles Darwin's test of survivability many times over in its first 50 years. Several times, over this period, the external as well as internal environments created adverse conditions; however, invariably the industry and its leaders demonstrated the ability to innovate and the willingness to change in order to survive and often come out better off from such experiences. Given the continuing spirit of innovation and the fast-maturing intrinsic strengths of this medium, there is every reason to be optimistic for the next 50 years and more.

With the practical ubiquity that the satellite technology now provides, it is no longer a mere cliché to say that every human being is now a potential customer for this medium. However, at no stage can we afford to forget that such customers have choices for what they need, as the competing media have also made impressive advances and often have very credible solutions to offer. Nevertheless, none of the other media can beat the ubiquitous solutions that genuinely customer-driven satellite systems have the potential to provide in ever-increasing situations. The responsibility for ensuring that the satellite systems are indeed customer driven does not belong only to the technologists, system designers, engineers, or sales and marketing executives, *but to all of them working in a truly interaction fashion*, totally focused on just one goal: what the customer needs, when he or she needs it, and at what price he or she is willing to pay. We close with the genuine hope that principles underscored and developed in this book will be helpful in this regard.

References

[1] Sachdev, D. K., "Historical Overview of the Intelsat System," *Journal of the British Interplanetary Society,* Vol. 43, 1990, pp. 331–338.

[2] Sachdev, D. K., " Three Growth Engines for Satellite Systems," *AIAA 20th International Communications Satellite Systems Conference,* Montreal, Canada, May 12–15, 2002.

[3] Maral, G., and M. Bousquet, *Satellite Communications Systems,* 4th ed., New York: John Wiley & Sons, 2002.

[4] Edelson, B. I., and W. L. Morgan, "Orbital Antenna Farms," *Astronautics and Aeronautics,* Vol. 15, September 1977, pp. 20–29.

[5] Iida, T., et al., "Communication Satellite R&D for Next 30 Years—Follow-On," *AIAA 20th International Communications Satellite Systems Conference,* Montreal, Canada, May 12–15, 2002, paper number AIAA-2002-1972.

[6] Gayrard, J., "Evolution of Telecommunication Payloads: The Necessity of New Technologies," *AIAA 20th International Communications Satellite Systems Conference,* Montreal, Canada, May 12–15, 2002.

[7] Dankberg, M., Acceptance Speech for the Satellite Executive of the Year, as reported in *Satellite News,* March 8, 2004.

[8] Fischer, R., "Transformational Communication Architecture," Keynote address, *AIAA 22nd International Communication Satellite Systems Conference,* Monterey, CA, May 9–12, 2004.

[9] Solman, F. J., et al., "The Ka-Band Systems of the Lincoln Experimental Satellites LES-8 and LES-9," *Proceedings of 7th AIAA Communication Satellite Systems Conference,* 1978, pp. 158–164.

[10] Sachdev, D. K., "Satellite Communication Technologies-Challenges for the 1980s," *Journal of Spacecraft and Rockets,* Vol. 18, March–April 1981, pp. 110–118.

[11] Draim, J., et al., "Beyond GEO—Using Elliptical Orbit Constellations to Multiply the Space Real Estate," *52nd International Astronautical Congress,* Toulouse, France, October 1–5, 2001.

[12] Cefola, P., et al., "The Orbit Perturbation Environment for the COBRA and COBRA Teardrop Elliptical Constellations," *AAS/AIAA Space Flight Mechanics Meeting,* San Antonio, TX, January 27–30, 2002.

[13] de Weck, O. L., et al., "Quantitative Assessment of Technology Infusion in Communication Satellite Constellations," *21st AIAA International Communications Satellite Systems Conference 2003,* paper number AIAA 2003-2355, 2003.

[14] McAlister, P., "Satellite 2-5 Statistics: Is Recovery a Mirage?" *Satellite 2004 Conference,* Washington, D.C., March 2–5, 2004.

[15] Chuberre, N., et al., "Satellite Digital Multimedia Broadcast System (SDMB) to optimize #GPP Architecture over Europe," *8th International Workshop on Signal Processing for Space Communications (SPSC 2003),* Catania, Italy, September 24–26, 2003.

[16] Fong, G. L., and K. Nour, "Broadband and Role of Satellite Services," Frost & Sullivan White Paper, 2004.

[17] Agnew, C., and S. Dutta, "Revitalizing the Mobile Industry," *International Satellite & Communication Expo (ISCe) 2003,* Long Beach, CA, August 2003.

About the Author

D. K. Sachdev is the founder and president of SpaceTel Consultancy, LLC, in Vienna, Virginia. This company provides business strategy and engineering support for satellite and wireless systems. Mr. Sachdev is also an adjunct professor at the George Mason University, Virginia, and teaches graduate courses in system engineering for telecommunication systems and project management.

From 1996 through 2000, as senior vice president, engineering and operations, at WorldSpace, Washington, D.C., Mr. Sachdev was responsible for the engineering, deployment, and operations of the first worldwide digital radio system. While at WorldSpace, Mr. Sachdev also contributed to the evolution of the XM Radio system and led the development of its system architecture and the initial stages of the engineering development of this system.

In recognition of his contributions, Mr. Sachdev received the Arthur C. Clarke Innovator's Award for 2003 "for his creative and innovative engineering prowess in the field of satellite communications for many years and culminating in the engineering implementation of the world's first operational direct broadcasting radio satellite system."

For almost two decades ending in 1996, Mr. Sachdev was at the center of the expansion of the INTELSAT's global telecommunication network. After establishing the in-house technology development team in the early 1980s, Mr. Sachdev led a team for the development, procurement, and deployment of 16 new satellites, now in operation for several years in the INTELSAT's and New Skies networks. Matching this effort in the space segment were several equally impressive efforts for international terrestrial networks.

Prior to crossing the oceans in 1978, Mr. Sachdev held several senior positions in the Indian Telecommunications Service and the associated industry.

Mr. Sachdev was a member of the founder team of the Telecommunication Research Center at New Delhi. He was in the core team for the development of microwave systems in India. During the early 1970s he created one of the largest design and development organizations in electronics and telecommunications at ITI in Bangalore. For his "outstanding contributions in electronics and telecommunications," Mr. Sachdev was awarded the prestigious Vikram Sarabhai Award in 1976.

D. K. and his wife, Usha, have lived in Virginia for the past 25 years. They have three children and six grandchildren.

Index

spacecraft payload, 168
system architecture, 167–70
total programming package, 168
See also Digital radio

XM Radio, 173–78
  availability, 177, 178

business-oriented approaches, 179
defined, 173
management, 179
spacecraft payload, 177
space segment, 174–77
system frequency plan, 177
See also Digital radio

For further information on these and other Artech House titles, including previously considered out-of-print books now available through our In-Print-Forever® (IPF®) program, contact:

Artech House
685 Canton Street
Norwood, MA 02062
Phone: 781-769-9750
Fax: 781-769-6334
e-mail: artech@artechhouse.com

Artech House
46 Gillingham Street
London SW1V 1AH UK
Phone: +44 (0)171-973-8077
Fax: +44 (0)171-630-0166
e-mail: artech-uk@artechhouse.com

Find us on the World Wide Web at:
www.artechhouse.com